Vanesa Magar

Sediment Transport and Morphodynamics Modelling for Coasts and Shallow Environments

Sediment Transport and Morphodynamics Modelling for Coasts and Shallow Environments

Vanesa Magar

CRC Press
Taylor & Francis Group
Boca Raton London New York

CRC Press is an imprint of the
Taylor & Francis Group, an **informa** business

CRC Press
Taylor & Francis Group
6000 Broken Sound Parkway NW, Suite 300
Boca Raton, FL 33487-2742

© 2020 by Taylor & Francis Group, LLC
CRC Press is an imprint of Taylor & Francis Group, an Informa business

No claim to original U.S. Government works

Printed on acid-free paper

International Standard Book Number-13: 978-1-4987-5346-3 (Hardback)

**Visit the Taylor & Francis Web site at
http://www.taylorandfrancis.com**

**and the CRC Press Web site at
http://www.crcpress.com**

'The supreme goodness is like water.
It benefits all things without contention.
In dwelling, it stays grounded.
In being, it flows to depths.
In expression, it is honest.
In confrontation, it stays gentle.
In governance it does not control.
In action, it aligns to timing.
It is content with its nature,
and therefore cannot be faulted'.
Tao Te Ching—passage no. 8

Contents

Preface

When Tony Moore from CRC Press came to see me back in 2011 and suggested I write a book, I had to ponder this for some time. After several months, I decided to send him a book proposal on sediment transport and morphodynamics modelling in coastal environments, to see whether he would agree. The book proposal was, in general, well received by Tony and the reviewers, but most importantly, Tony was confident I could undertake this project to completion. I invited a senior researcher, Prof. Alan G. Davies, with whom I had collaborated during the SANDPIT project (EC Framework V Project, contract no. EVK3-2001-00056), to be a co-author. But Alan was also confident I could undertake this without much additional input. Thus, in 2015, I signed the contract with CRC Press and embarked on this adventure. Initially, Tony was giving a lot of advice to help me organize my time so that I would finish in two years. However, it took a bit longer than that, not by lack of interest but lack of time. In 2015, I had moved back to Mexico and was involved in several renewable energy projects. Our work required a certain level of social activism, with some focus on marine spatial planning and management, while trying to influence national policies. Such a turn in my career was due to a realization that we need to take urgent climate change action. We are endangering all forms of life with our anthropogenic activities and greenhouse gas emissions. We need, therefore, to focus on projects of significant benefit for the Planet and our environment. Some of these worries affected my writing of this book. Also, they influenced the selection of pressing issues to be addressed, presented in the final chapters. I have, however, been relatively loyal to the original content plan.

Although this book is mostly a research monograph, I hope it is of interest to researchers and practitioners alike. A strong focus is on the dynamics of coastal environments at different scales and under a variety of forcings, using models, *in-situ* data, and remote sensing techniques. A broad overview of the different chapters is as follows. Chapter 1 briefly defines different spatio-temporal scales, as well as several essential terms and concepts. We present linkages with modern observation tools and regional models and introduce definitions and concepts related to coastal management. These concepts and tools may be used to solve endless research questions and provide interesting discussions. They are essential not only for the analysis of the physics and the dynamics, but also for assessing biogeochemical and ecological problems and for communication with colleagues, authorities, or other interested parties. Chapter 2 provides essential information on the physics of open coasts and sheltered environments, including grain size distributions, sediment composition, bottom boundary layer dynamics, wave and tidal dynamics, and their effects on sediment transport and morphodynamics modelling, the main focus of this work. Chapter 3 then introduces basic and advanced regional modelling tools. I attempt here to guide you to best practices, for example, on grid generation, bathymetry interpolations, methods of solution. The chapter also covers different types of modelling approaches in wave-dominated and in sheltered environments. Data-driven methods are also discussed, as such decompositions into a small number of dominant patterns can be very useful to understand long-term dynamics. Finally, Chapter 4 provides detailed descriptions of typical applications of regional models, including estuarine dynamics, shoreline evolution, and environmental impacts of offshore structures.

Vanesa Magar,
Ensenada, Baja California, Mexico, 18 July 2019.

Acknowledgements

I would like to thank those who have made this project possible. First, thanks to CRC Press, whose support has been instrumental. I was glad to have them touching base from time to time and checking whether the draft was (or not) making any progress. As the bulk of this book took shape, I benefited significantly from Alan's thorough reviews, in particular, his comments about adding examples and important results. Thank you, Alan, for all the time you invested in the past year to this project. I learnt how to synthesize well essential results after I read the book edited by Reginald Uncles and Steven Mitchell; that book was extremely valuable during the revisions of this work. I amply recommend it to anyone interested in these topics.

Thanks to Anahi Bermudez Romero and Victor Manuel Godinez Sandoval for their help with the figures and schematics reproduced in the text. Thank you as well for looking after important deliverables in other projects, while I worked on this book. Finally, thanks to my husband and co-leader of the GEMlab, Markus Gross, and to our son Damian Gross-Magar, for their patience.

Dedicated to those pursuing greater purpose and goodness in life, despite hostilities and pressures.

Author

Vanesa Magar is a senior researcher and associate professor at the Centro de Investigación Científica y de Educación Superior de Ensenada in Baja California, Mexico. She was formerly a researcher and then a lecturer at Plymouth University, UK.

1

Introduction

Coasts and shallow coastal environments evolve in response to different forcings and constraints, for example, the supply of sediment; the action of winds, waves, currents, and density gradients; the effects of extreme events such as hurricanes, tsunamis, and landslides; the presence of beaches, dunes, coastal lagoons, estuaries, or deltas; and the changes induced by human intervention and global warming on the coast. These forcings and constraints act at different space scales and timescales. Large spatio-temporal scales provide the boundary conditions for a model that can generate the small spatio-temporal scales, and, in turn, the physics resolved at small spatio-temporal scales are parameterised in models at large spatio-temporal scales (Brommer & Bochev-Van der Burgh 2009). Thus, there is an exchange of matter, momentum, and energy between scales, known as the *cascade hierarchy* (Cowell et al. 2003). Moreover, processes and bedforms at the same spatio-temporal scale are linked to one another through dynamic interactions, which is known as the *primary scale relationship* (de Vriend 1991). A schematic of the cascade hierarchy is shown in Figure 1.1. The space scales and timescales shown are in orders of magnitude only, implying that there is some flexibility on the boundary limits for the scale boxes shown, depending on the circumstances, for example, under extreme events or long recovery time periods. We will now discuss in more detail the primary scale relationships in the context of coastal geomorphology and dynamics or coastal morphodynamics.

1.1 Physics and Modelling at Different Scales

The study of coastal morphodynamics involves investigating the changes to the physical processes and bedforms in the coastal environment over a broad range of scales in space and time, from the microscale to the macroscale. The motions near the bed are crucial as the bed is one of the most important sources of sediment, together with bedform migration and estuarine discharges. Sediments transported by the flow may be divided into three different transport types depending on their median grain diameter and their dynamics: bedload, suspended load, and wash load. The differences between these three types of sediment load will be explained in Section 1.1.1 on microscale

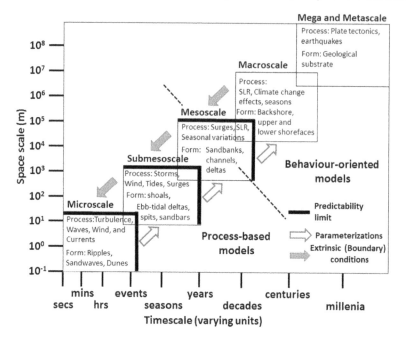

FIGURE 1.1

Cascade hierarchy of spatio-temporal scales. (After Brommer & Bochev-Van der Burgh 2009.)

processes. The types and concentrations of the sediments will strongly depend on the coastal environment, which may be classified into three simple types: sandy or pebbled coasts, erodible or nonerodible bluffs and cliffs, or sheltered low-lying lands and estuaries. Such environments may extend from tens, to hundreds, to thousands of meters and therefore may be affected by submesoscale, mesoscale, or macroscale processes. The US Federal Emergency Management Agency (FEMA) defines six different beach settings, representative of the settings generally found on the Pacific coast of the US, according to their morphology and vulnerability to storms (FEMA 2015).

1. Sandy beaches backed by low sand berms or high sand dune formations

2. Sandy beaches backed by shore protection structures

3. Cobbled, gravelly, shingle, or mixed-grain-sized beaches and berms

4. Erodible coastal bluffs

5. Nonerodible coastal bluffs or cliffs

6. Tidal flats, coastal lagoons, wetlands, and other reduced energy basins.

Beach settings 1–3 correspond to open beaches, beach settings 4 and 5 correspond to bluffs and cliffs, and beach setting 6 correspond to low-lying sheltered environments. The extent of the study domain and the resolution at which one is carrying out a study clearly affect the scale of the processes that need to be considered. For consistency, it is therefore very useful to identify the morphological and dynamical mechanisms that affect the evolution of different coastal environments at different scales.

1.1.1 Microscale

The bottom boundary layer of the nearshore region is governed by microscale processes and bedforms. At the very low end of the microscale, at length-scales below $O(0.1)$ m and timescales of the order of seconds, turbulence and individual particle motions are the dominant processes. Bedload differs from suspended load in that bedload sediment particles remain close to the seabed and move by traction or saltation. Traction occurs when the sediment and the fluid have comparable densities, while saltation is common when the particle density is large in comparison to that of the surrounding fluid (Charru et al. 2013). When studying bedload transport, it is important to consider the threshold of motion of the sediment, which occurs at Shields number, θ, equal to the critical Shields number, θ_c. The Shields number is a dimensionless shear stress and depends on the geometry of the grains (whether they are smooth or rough), the fluid viscosity, the fluid and the particle densities, the median grain diameter, and the friction velocity at the seabed. The threshold of motion may also be defined in terms of excess pressure gradients, through the Sleath parameter (Sleath 1982). More formal definitions of the boundary shear stress, the Sleath parameter, and the boundary Reynolds number will be provided in Chapter 2. Once the shear stresses generate a lifting force that is large in comparison to the particles' stabilizing forces, bedload sediments start rolling or hopping on the bed, moving by traction or saltation. Once turbulent forces are large enough to lift the sediment higher into the water column, the bedload turns into suspended load, and the particles are advected by the main flow. This occurs when the fluid velocity fluctuations become comparable, in magnitude, to the sediment settling velocity. In other words, the balance between erosion and deposition depends on the flow speed near the bed and the sediment median grain diameter (Hjulström 1935). Materials in suspension are maintained as suspended load by turbulent motion (Bagnold 1966); they may have come from upstream reaches, carried downstream as wash load. Wash load sediments are fine particles such as clays, silts, or very fine sands, with very small or negligible settling velocities, and are not found in significant quantities at the bed (Einstein 1950). These particles can be carried by the flow at lower flow velocities than coarser materials (Hjulström 1935). It is difficult to identify the boundary between bedload and suspended load, as hopping grains slowly transition into grains carried by the flow (Bartholdy et al. 2008). The bedload component of sediment transport is the dominant component in bedform migration, but suspended sediments also have a contribution in bedform evolution processes. Neither bedload nor suspended load transport models, however, explicitly determine bedform changes, which have to be determined using Exner's equation (Exner 1920, 1925)—an equation of conservation that balances sea-level changes with sediment flux gradients. Felix Exner was the first to express this balance quantitatively, thus becoming the father of morphodynamics. More details on sediment transport formulations will be presented in Chapter 2.

In the instantaneous swash zone, a sheet flow layer forms as a result of the interaction between lateral pressure gradients caused by the swash bore or wave set-up gradients, shear stresses on the water surface, the gravitational force, and the undertow (granular) flow. Sheet flows are multiphase, fluid–sediment mixtures with a very high concentration of sediment. Sheet flow layers occur not only in the swash zone, but also in the surf zone, near the seabed, under strong bottom streaming or wave acceleration skewness (Nielsen 2006, Van der A et al. 2010).

The microscale generally covers lengthscales of O(0.1–100) m and timescales of O(0.01–1) days. The processes of relevance include turbulence, waves, winds, and nearshore currents, all acting to move sediments in the backshore, in the upper shoreface, or between these two regions (Dean & Dalrymple 1984). The backshore is the region of the beachface-seabed that is above the run-up limit, while the upper shoreface is the region between the run-up limit and the depth of wave influence. The forcings in the bottom boundary layer are responsible for the evolution of small-scale bedforms, such as ripples. Ripples are the smallest bedforms found in unidirectional and oscillatory flows, with typical heights of the order of a few centimeters. Within the microscale range, one may also consider 100 m large river megadunes (Charru et al. 2013). Under unidirectional flows, all bedforms migrate significant distances downstream, but under oscillatory flows, migration usually occurs when there is a net flow, either onshore or offshore (Van der Werf et al. 2008). Figure 1.2 shows sheet flow and rippled bed flow schematics. In the sheet flow regime, the fluid–sediment dynamics of a slab of mixture, as depicted in Figure 1.2, depend on the balance between the pressure gradients, on the left and right walls; the shear stresses at the top boundary; the granular friction at the bottom boundary; and the downslope gravity within the mixture. The picture on the top left shows some artificial grains in a sheet flow experiment, while the picture on the bottom left shows a naturally rippled bed on an exposed tidal flat (from Clapham 2015). In the rippled bed regime, a similar balance between forces needs to be considered. However, the flow above rippled beds gets separated from the bed at the brink point, or point of flow separation, when the flow moves from the stoss side (on the upstream face) to the lee side (on the downstream face) of the ripples. Coherent vortices laden with sediment form on the lee side. Some of the sediment is ejected with the flow at the point of flow separation on the ripple crest before the vortices form, but the vortices also cause erosion on the stoss

Sheet flow and rippled bed flow dynamics

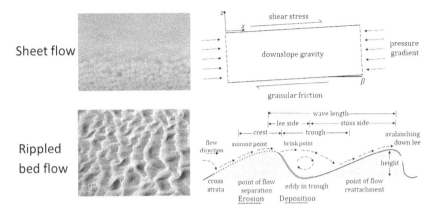

FIGURE 1.2
Sheet flow and rippled bed flow schematics. Sheet flow picture after Clapham (2015), sheet flow diagram after Lanckriet & Puleo (2015), and rippled flow picture and diagram after Clapham (2015).

side of the following ripple on the downstream direction, setting some bedload in motion. When the flow changes direction, the vortices are ejected up into the water column, carrying sediment with them. The process is described in detail in many publications on wave dynamics and sediment transport above rippled beds—see, for example, Malarkey et al. (2015), Van der Werf et al. (2008), and Malarkey & Davies (2002).

Other bedforms, such as sandbars or sandwaves, may migrate offshore under the action of sea storms and migrate shorewards during calmer sea state conditions. Nearshore currents may have a significant longshore component, or *longshore drift*, that causes sediments to move alongshore rather than cross-shore. As Cowell et al. (2003) highlight, the nearshore waves and currents drive local longshore and cross-shore *sediment budgets*, i.e. sediment gains and losses, and define the position of the shoreline.

1.1.2 Submesoscale

Submesoscale processes are, by definition, all processes affecting the dynamics between the largest turbulent scales and the smallest mesoscales (Hoover et al. 2015). Open beaches and coastlines are affected by submesoscale processes at scales of days to years and lengthscales of O(100) m to O(1) km. A detailed cross-shore schematic of open, sandy, nearshore regions, including the backshore and the upper shoreface, is shown in Figure 1.3. The upper shoreface and lower shoreface extend to the limit of the continental shelf, usually found at a depth of around 200 m, where the continental shelf is in free interaction with the deep ocean.

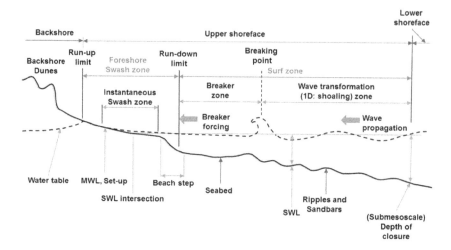

FIGURE 1.3
Beach schematic. (After Elfrink et al. 2006 and Chardón-Maldonado et al. 2015.)

The depth of closure for open beaches is the depth beyond which the envelope of variation of profile measurements stops changing. The depth of closure depends, mostly, on waves, currents, tidal forcings, and internal system dynamics such as bedform migration. The depth of closure (shown in Figure 1.3 for episodic events to seasonal scales) was investigated during the MAST-III project 'Predicting aggregated-scale coastal evolution' funded by the EU (Nicholls, Birkemeier & Hallermeier 1996, Nicholls, Larson, Capobianco & Birkemeier 2001). In that project, it was defined as the depth beyond which there is no significant seabed change or sediment exchanges between the upper and the lower shoreface (Morang & Birkemeier 2005). The term 'significant', however, is very subjective, and the depth of closure is, rather, a conceptual limit between two regions that respond, in a morphodynamic sense, at different spatio-temporal scales (Nichols & Biggs 1985). For example, the depth of closure shown in Figure 1.3 responds to submesoscale and mesoscale processes. A process-based modelling approach is commonly used at these scales, except when analysing long-term morphodynamics at well-surveyed sites, where a data-driven approach can be implemented. The lower shoreface boundary, in contrast, is a macroscale–megascale depth of closure, and at these scales, it is more common, rather, to use behavioural models such as the one by Cowell et al. (2003). The characteristic dynamics at the macroscale–megascale boundary are outside the scope of this book and so are, to some extent, the microscale dynamics. The focus of this book is mainly on the behaviour from the submesoscales to the macroscales.

At submesoscales, the Rossby and Richardson numbers generally are close to 1, which has significant implications for ocean mixing and primary productivity (Thomas et al. 2013). Submesoscale oceanographic processes include density fronts and filaments, wakes generated by flow–bottom interactions (topographic wakes), and surface and internal persistent coherent vortices (McWilliams 2016). In deeper waters, submesoscale processes become more important in regions where there are large temperature gradients over short distances, giving rise to large vertical velocities, as the colder waters try to flow under warmer waters. High-resolution remote sensing observational images, such as that shown in Figure 1.4, of the heat and suspended matter plume caused by a underwater volcano eruption, taken using Moderate Resolution Imaging Spectroradiometers (MODISs), Medium Resolution Imaging Spectrometers (MERISs), 0.46m panchromatic (black and white) mono and stereo satellite image data from DigitalGlobe's WorldView-2 satellites, or Advanced Spaceborne Thermal Emission and Reflection Radiometers (ASTERs), suggest that modern remote sensing technologies are appropriate tools to analyse the submesoscale dynamics of natural hazards and plankton blooms, based on the horizontal-scale resolution of filamentary patterns, small eddies, and the complexity of the mixing processes. The ASTER RGB image on the left in Figure 1.4 is from ASTER and has a spatial resolution of 15–90 m; it covers an area of 48.6 by 57 km and is located near 27.6° N latitude, 18.1° W longitude. The MODIS–MERIS diffuse attenuation

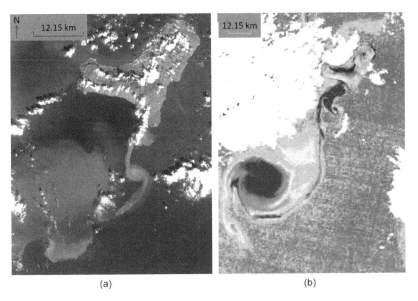

(a)   (b)

FIGURE 1.4
Images of the heat and suspended matter plume caused by the 'El Hierro' underwater volcano eruption obtained with satellite remote sensing technologies. (a) ASTER RGB image on 1 November 2011. (Image source—reproduced with permission: www.jpl.nasa.gov/spaceimages/details.php?id=PIA15017.) (b) MERIS $K_d(490)$ image on 4 November 2011. (From Marcello et al. 2015—reproduced with permission.)

coefficient, or K_d, at the wavelength of 490 nm, or $K_d(490)$, shown on the right has a spatial scale about 2.3 times smaller than that of the ASTER RGB image on the left, and thus covers an area that is roughly 2.3 times larger; it was taken 3 days later and shows how the mesoscale eddy generated by the plume of suspended matter evolved and moved several kilometres to the south-west. Segmentation algorithms to detect the initial structures, followed by advanced structure growth algorithms, permit the separation of submesoscale from mesoscale patterns and the study of the filament-like and front-like submesoscale patterns observed in the images (Marcello et al. 2015).

High-resolution remote sensing technologies for water quality research provide unprecedented information on submesoscale suspended matter horizontal dynamics, not only for natural but also for anthropogenic hazards. Xian & Weng (2015) discuss interesting remote sensing techniques for water quality applications. Inland and coastal water environments suffer from direct impacts of increasing urban, industrial, or agricultural land use, either from point or nonpoint sources. Such land use changes modify the characteristics of local and regional runoff flows, because the construction of impervious surfaces does not allow water to penetrate the soil, causing coastal runoffs into the sea with larger discharges or causing discharges of water with higher pollutant and nutrient concentrations. Therefore, it is crucial to develop effective water quality monitoring and assessment methods for sustainable water resource management. This assessment depends on accurate, intensive, and long-term water quality data acquisition. However, traditional water quality evaluation has been too limited on temporal and spatial scales to address factors affecting, for example, harmful algal blooms or other types of water quality pollutants such as total suspended sediments (which affect turbidity), chemicals, dissolved organic matter, thermal releases, or aquatic vascular plants. Remote sensing techniques provide a spatial and temporal coverage that is impossible to obtain from *in-situ* measurements. Some substances can change the backscattering characteristics of surface waters, and remote sensing techniques can identify these features through the spectral signature backscattered from the water, usually in the visible and near-infrared (NIR) part of the electromagnetic spectrum (Matthews 2011). Then, the measured backscatter signal can be related to some water quality parameter using either empirical or semianalytical methods (Xian & Weng 2015). Most water quality parameters, however, can only be inferred indirectly from measurements of other water quality parameters (Ritchie et al. 2003, Matthews 2011). The parameters that can be derived from remotely sensed data include phytoplankton pigments such as chlorophyll (Chl-a)(Yacobi et al. 1995, Brivio et al. 2001, Giardino et al. 2005, Gitelson et al. 2009), total suspended sediment (TSS) concentration (Onderka & Pekarova 2008, Doxaran et al. 2009, Oyama et al. 2009), absorption by CDOM (Kutser et al. 2005), water clarity (Olmanson et al. 2008), turbidity, and water temperature (Matthews 2011), to name but a few. The interested reader is referred to Xian & Weng (2015) and references therein for more details on how these parameters are determined.

In submesoscale regions, the vertical motions are also important for transport and mixing of suspended matter up and down the water column. For planktonic cells, this means they can reach nutrient-rich, light-rich surface waters. In coastal areas, plankton blooms and tracer dispersion at submesoscales are accelerated by cold-water upwelling, shore-perpendicular surface currents, and warm-water downwelling driven by surface wind shear stresses and Ekman circulation. Despite technological advances, submesoscale processes are still difficult to model or observe. In most cases, it is necessary to either have *in-situ* data available or plan bespoke fieldwork campaigns to generate fit-for-purpose observations. Temperature and salinity are two crucial oceanographic variables needed to analyse submesoscale ocean dynamics in waters where baroclinic instabilities are an important driver of the dynamics. However, for coastal hydrodynamics in shallow environments, the submesoscale processes of most interest associated with temperature and salinity gradients correspond to estuarine regions or coastal rain runoff sites, where the interaction and mixing of freshwater and saline ocean waters at different temperatures occurs. From a dynamical perspective, analyses of the high-vorticity boundary layers generated by the interaction of ocean currents and topography are also of interest (Srinivasan et al. 2017). Topographic wakes involve separating drag-generated boundary layers (BLs) (Molemaker et al. 2015), either from the bottom (BBLs) or the side (SBLs) of the topographic domain, and its subsequent roll-up through barotropic–centrifugal instabilities. These current–bottom interactions can inject a significant number of submesoscale coherent vortices (SCVs) into the ocean interior. As explained by Srinivasan et al. (2017), the strength of these SCVs is proportional to the bottom current speed and the local topography's slope. The separating bottom shear layers have horizontal scales of $\sim$100m and vortex Rossby numbers larger than 1. The Rossby number is a dimensionless number that determines the balance between inertial and Coriolis forces,

$$Ro = \frac{U}{fL},$$ (1.1)

where U is some characteristic speed (here a speed of the vortex, e.g., its mean speed of propagation), L a characteristic length (here a characteristic length-scale of the vortex, e.g., its average diameter), and f is the Coriolis coefficient. In fact, near-equatorial SCVs can have larger vortex Rossby numbers (of order $\sim$20) that are also accompanied by significant interior dissipation and mixing. Therefore, the boundary layer separation generates an instability of filaments with Rossby numbers of order $\sim$1, which grow through an upscale process into SCVs or even mesoscale patterns.

Suspended sediment concentrations, on the other hand, depend mostly on hydrodynamic processes such as tides, density gradients, and estuarine discharges, together with sediment availability, sediment size, and flocculation (Chernetsky et al. 2010). These processes generate several estuarine turbidity maxima (ETMs) along the estuary (Lin & Kuo 2001). At these maxima, oxygen concentration is really low (Talke et al. 2009), and the hypoxic conditions may cause local damage to aerobic organisms; this may have a negative impact on all trophic levels. From the physical process modelling perspective, in estuarine settings it is common to assume width-averaged tides varying vertically and longitudinally—i.e., along the transversal direction of the estuary (Chernetsky et al. 2010). Width-averaged models are common in industry; for example, SOBEK (www.deltares.nl/en/software/sobek/), developed by Deltares, is one of such models, which is now integrated into the Deltares Flexible Mesh suite (www.deltares.nl/en/software/delft3d-flexible-mesh-suite/). However, as highlighted by MacCready & Geyer (2010), tides are not the only relevant hydrodynamic process in estuaries. Tidally averaged circulation, mixing and sediment fluxes, and other recent developments have emphasised the importance of eddy covariance–shear covariance in the generation of the exchange flow (Dijkstra et al. 2017), also known as the estuarine circulation. Even in weakly stratified flows, vertical and lateral advection terms are significant drivers of the mixing process (MacCready & Geyer 2014). Moreover, it is known that the embankments and the complex topography of estuaries cause additional friction and slow down the tidal currents. Hence, while tidal elevations may be assumed 1D along the estuary, velocity magnitude differences between the sides and the middle of the estuary show clear evidence of shearing and 3D effects on the tidal velocities (Li & Valle-Levinson 1999, Waterhouse et al. 2011).

From the morphodynamics perspective, submesoscale bedforms include sandwaves, coastal sandbars, megaripples, or large riverine or estuarine shoals such as ebb and flood delta shoals. Sediment transport and morphodynamic processes in estuaries and coastal lagoon systems may evolve significantly at space scales and timescales typical of the submesoscale. This is because coastal settings may vary considerably within the same region. In Figure 1.5,

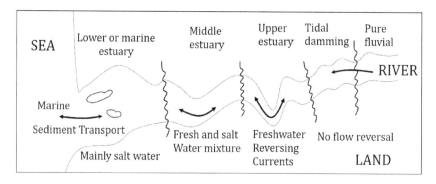

FIGURE 1.5

Estuary schematics showing the different estuarine regions depending on the flow direction and water salinity. (After Perillo 1995 and Syvitski et al. 2005.)

for example, we show a schematic of the three different reaches along an estuary, based on the interplay between marine and fluvial dynamics (Perillo & Piccolo 2011). In estuarine environments, silts, clays, and organic matter coming from the river form bedforms that are more resistant to shear stresses and motion. The three estuarine settings are the marine estuary, where the estuary connects with the sea and where the influence of waves and tides is important; the middle estuary, corresponding to the region with largest mixing of fresh and salt water; and the fluvial estuary, where the water is fresh but the tides still have a dynamical influence. The marine estuary is the location where largely marine sediments are found, with sandbanks and shoals forming and migrating in response, mostly, to tidal forcings, but where storm waves may affect sediment transport and seabed evolution, in particular under extreme events. The middle estuary may be well mixed or stratified. In classical estuaries, the monthly circulation depends on the freshwater input, and as salt water is heavier and denser than freshwater, usually there is a salt water inflow at the bottom and freshwater outflow at the surface. Inverse estuaries, on the contrary, are estuaries where there is little or no freshwater inflow, and strong surface evaporation causes the water in the surface layers to become saltier than the water in the bottom layers, so the ocean water flows into the estuary, becomes saltier, and then flows outwards close to the estuary's bottom (Valle-Levinson 2010).

1.1.3 Mesoscale

As the timescale increases to O(1–100) years, and space scales to O(10–100) km, also the number of processes affecting the depth of closure increases. The Rossby number, Ro, as defined in Equation eq. (1.1), and the Richardson number, which in oceanography is defined in terms of the Brunt–Väisälä frequency, N, also known as the buoyancy frequency, and a characteristic flow shear measure, dU/dz,

$$Ri = \frac{N^2}{(dU/dz)^2},\qquad(1.2)$$

both play an important role at these scales, where the ocean dynamics are characterised by $Ro \ll 1$ and $Ri \gg 1$, respectively (Thomas et al. 2013). N is the angular frequency at which a vertically displaced parcel will oscillate within a statically stable environment and is a measure of ocean stratification. Ri is a measure of mechanical (through the shear stress term) and density (through the buoyancy frequency) effects on the water column. When Ri is large, as in mesoscales, turbulent mixing across stratification layers is generally suppressed.

In relation to sediment transport and morphodynamics modelling, the morphological evolution of shallow coastal regions is dominated by long-term tides and event-driven storm surges, sea-level rise, and seasonal variations, but it is the large bedforms such as deltas and sandbanks that migrate under the hydrodynamical forcings. The temporal scales of interest range from months to years and decades. These scales are relevant to coastal researchers interested in the impact of climate change on the coast. Process-based modelling may still be implemented but only if the codes are highly optimised or run in clusters of high-performance computers. However, due to the computational costs, and the uncertainties associated with the accuracy of process-based models at such scales, it is common to adopt instead behavioural, hybrid, or reduced complexity approaches, such as that presented by Reeve et al. (2016), for example. Data-driven models, when sufficient data is available, are an excellent alternative to behavioural approaches. In some cases, the best approach may be a combination of different methodologies. Despite the progress achieved in the last 20 years, there are many remaining questions to be addressed whereby our understanding of coastal morphodynamics can be improved.

In a full numerical modelling set-up, the meteorological, hydrodynamical, and morphological equations are coupled and feedbacks affecting the evolution of these three systems are considered in the conservation Equations (Warner et al. 2008). However, in long-term simulations, such models are very costly, and hence a form of reduction of the forcing conditions is applied. On open coasts, this reduction can be to the wave forcings, but in estuaries, it is usually to the tidal forcings. Mesoscale evolution of an estuary may be assessed with such representative tides for the site of interest (Winter 2006). This representative tide is the one that best reproduces the observed morphological dynamics, in terms of erosion and deposition patterns, over yearly timescales. A reduction in timescales between the hydrodynamic and the morphodynamic models also leads to further savings in computational costs. Both the simplified forcings and hydrodynamic–morphodynamic timescale factors need to be calibrated so that the model predictions stay reliable. This may be difficult to achieve without a series of echosound bathymetries obtained over a long period of time, for the region of interest. In order to compare the models with observations, some processing of the bathymetric data may be necessary, in order to separate the bathymetric patterns into primary (largest), secondary, tertiary, or even smaller bedforms, depending on the spatial scales of the study (Winter 2011).

A correct model set-up requires information on the geometry of the coast and the channels in the model domain; tidal ranges and tidal principal component characteristics; the catchment area, its mean tidal volume, and its mean discharge; the mean grain size of the sediment bed and the variations in grain size between channels, tidal flats, and embankments, as well as variation between the estuary's low and high reaches; suspended sediment characteristics, such as background (mean), maximal, and minimal suspended sediment concentrations during a tidal cycle; and bedform location, type, size, sediment composition and evolution, if known (Winter 2006). Many mesoscale studies focus on long-term dynamics, where a good knowledge of the history of human interventions is needed, together with the observed dynamical changes that such interventions caused in the system. Subaqueous bedforms at large to small scales are important roughness elements interacting with the flow, changing both the direction and the magnitude of the current velocities. These flow–bathymetry interactions may be analysed with the help of numerical models, but over long periods of time, as mentioned before, models with reduced dynamics are preferable to more detailed models.

When the bathymetry has been surveyed over long periods of time, it is possible to apply data-driven approaches of bedform evolution. Such methods do not require any information on the forcing mechanisms, as can be seen in the study of Reeve et al. (2008), for example. However, it is known that storms influence nearshore dynamics (Magar et al. 2012), and cyclic morphodynamic patterns may be related to cyclic long-term forcings, such as the North Atlantic Oscillation (Magar et al. 2012) or the El Niño Southern Oscillation (Barnard et al. 2015).

1.1.4   Macroscale

The macroscale covers timescales of O(100–1000) years, and space scales of O(100–1000) km. Coastal behaviour at these scales has been studied in detail by a number of previous researchers specializing in behaviour-based modelling (Brommer & Bochev-Van der Burgh 2009, Cowell et al. 2003, and references therein). These are long-term, large-scale models covering the whole shoreface (backshore and foreshore) and involving simplified physical assumptions. For long-term mesoscale modelling of sites that have been intensely monitored over long periods of time, morphodynamic evolution can also be studied with data-driven models; however, usually such surveys cover only a small area of the shoreface. For example, the surveys used to analyse sandbank evolution in Great Yarmouth (see Reeve et al. 2008, for further details) cover an area of 200 km². Coastal tract models, involving the backshore and the upper and lower shoreface, necessarily need to be based on simpler, hybrid models (Reeve et al. 2016). Shoreline evolution associated with changes in the continental shelf and the coastal plain are of low order (Stive 2003) in the cascade hierarchy defined in Figure 1.1. Knowledge of shoreline evolution is of crucial importance for coastal management and protection, and the lateral displacements of the shoreface, determined through morphodynamic coupling mechanisms between the shoreface and the backshore determine shoreline advance and retreat trends. Behaviour-oriented models are commonly used at these scales.

A behaviour-based model may be defined as a phenomenological model, based on simplified semiempirical averaged formulae (de Vriend et al. 1993, Niedoroda et al. 1995). A behaviour-based model formulation is schematised in Figure 1.6, showing the different steps involved in simplified modelling approaches. It is worth noting that similar schematics may describe a process-based model, except for the boundary conditions and the parameter estimations. A behaviour-based model may initially use bathymetry inputs from observations or an explicit form of the bathymetry as predicted from a

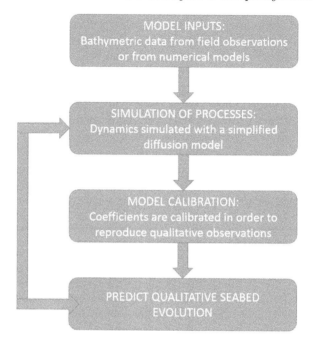

FIGURE 1.6
Procedure protocols for behaviour-based modelling. (After Magar 2008.)

process-based model. The bathymetry is then updated through a diffusion-type equation, simulating seabed volume or seabed height changes, balanced by cross-shore and longshore sediment fluxes. The resulting bathymetry is then compared with observations, and the model parameters are adjusted until the level of qualitative agreement between model predictions and observations is acceptable. In the diffusion-type model, assumptions are made on the physical processes that dominate the long-term dynamics of the system. For instance, in one-line and N-line models in general (Hanson & Larson 2000, Falqués 2003, Falqués & Calvete 2004), it is assumed that the shoreline or the isobaths are in equilibrium and that diffusion is the dominant process. In such a model, the diffusion coefficient is adjusted until the model reproduces the behaviour to a required accuracy. Many behaviour-based models are numerical in nature due to the complexity of measured boundary conditions.

As the focus of this book is on higher-order modelling, from the mesoscale down to the microscale, we will stop here with the description of the macroscale and will not talk about lower-order methods at all. Hence, we move on to a very important topic, in particular for determining climate change mitigation strategies and for their implementation: coastal management practice.

1.2   Coastal Science and Coastal Management

In addition to the curiosity-driven research that is essential to improve our understanding of coastal system dynamics, applied researchers also engage in end-user-driven research, in which coastal managers pose a question for coastal scientists and engineers to answer and report the solution back to them, following what Van Rijn et al. (2005) called 'traditional communication'. In more 'innovative' forms of communication, however, coastal managers, scientists, and engineers identify coastal state indicators and form hypothesis networks to answer the question together, through a joint collaboration.

The methodology followed in innovative end user–researcher communication strategies is a three-step approach illustrated in Figure 1.7. The most important concept introduced with the strategy is the 'coastal state indicator' (CSI) concept, which identifies the variables or parameters that are most relevant for the end user question. Ideally, CSIs should be determined in close collaboration with end users, either through a joint project or through an interactive consultancy process. The joint identification of essential CSIs provides a framework, a 'basic' frame of reference, shared by specialists and end users, which helps improve communication and ensures a shared perspective on the effectiveness of the solutions the specialists propose (Van Koningsveld et al. 2005). This frame of reference consists of the collection of essential CSIs, or elements, for effective communication of the problem and the associated solutions that were implemented. The frame-of-reference methodology may be applied to coastal policy and management. The choice of essential CSI is

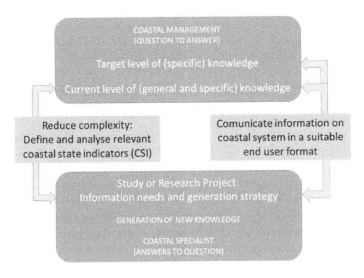

FIGURE 1.7
'Innovative' communication strategy. (Based on Van Rijn et al. 2005 and Van Koningsveld et al. 2005.)

mostly driven not only by the given problem, but also by practical limitations or technology availability. CSIs may be derived from video or satellite imagery, from oceanographic and coastal measurements, from laboratory observations, or from numerical models.

A generic approach for identifying essential CSIs, based either on the needs of short-, medium-, and long-term management practice and innovation was proposed by Van Koningsveld et al. (2005). As a first step, 'problem-driven' CSIs are chosen within a management context, and key issues are identified and posed as questions that coastal managers need to ask about their site. Then, as a second step, each field site is assessed scientifically using a systems-driven approach, to define a set of 'systems-driven' CSIs; these are indicators, or predictors, of known dynamic elements of the coastal system. If an emerging new technology is used during the scientific exploration of the site, then such techniques may lead to novel predictors describing the site dynamics. The problem-driven and system-driven CSIs should be comparable. As a third step, CSIs are selected depending on whether they help end users address a particular associated question. The final set of CSIs is reported and presented to coastal managers in an end-user-friendly format. The process requires a series of end user–specialist discussions before an agreement is reached.

1.3   How This Book Is Organised

As mentioned above, there are still important questions to address on sediment transport and coastal morphodynamic modelling in nearshore environments, and some of them will become evident throughout this book. This monograph will provide a comprehensive synthesis of state-of-the-art research and development.

Chapter 2 covers the fundamental hydrodynamics of estuarine and coastal environments, properties of seafloor and estuarine composition, and hydroenvironmental interactions. Emphasis is placed on the feedbacks between small- and large-scale processes, and between short and large evolution timescales. The theory is placed within the context of current and emerging soft engineering approaches for coastal management and protection, in particular on managed realignment, coastal protection, and climate change.

Chapter 3 follows with an in-depth coverage of numerical modelling techniques, where the challenges imposed by the modelling at multiple scales and the coupling of the physics with the dynamics is emphasised. Detail on how to configure regional models for typical and advanced coastal engineering applications is provided. This book covers a wide range of environments, but particularly those for which shallow-water theory is applicable. Coupling of the models with surface waves is discussed, and case studies where waves are of importance are included.

Chapter 4 focuses on current and future model applications with an emphasis on climate change. Topics covered include storm surge and inundation modelling, shoreline management and coastal erosion, the Sand Engine case study of morphodynamics modelling for beach renourishment schemes, and a closing section on offshore renewable energy.

2

Fundamental Physics

2.1 Threshold of Motion

Erodible beds in seashores and river deltas provide the necessary sediments for natural coastal accretion throughout the world. Such erodible beds may have different compositions and percentage content of different sediment sizes, and these sizes and material types in turn determine the degree of cohesiveness or noncohesiveness of the bed. A typical coarse (i.e., nondetailed) sediment size classification scheme is shown in Figure 2.1, where the sediments have been divided into only three types: mud (which may be divided into clay and silt), sand, and gravel (Van Maren 2009). The nominal particle sizes shown in the diagram follow Wentworth's grade scale. With these three types, bed compositions can be classified into fifteen major, three-way textural sediment groups, which can be very clearly represented in the form of a sectioned triangle, as shown in Figure 2.2. Another sectioned triangle may be defined for clay–silt–sand mixtures, as shown in Figure 2.3. The textural classification of sand–silt–clay shows that mud has a silt:clay fraction between 2:1 and 1:2, while silts have at least a 2:1 silt:clay fraction and clays at least a 1:2 silt:clay fraction.

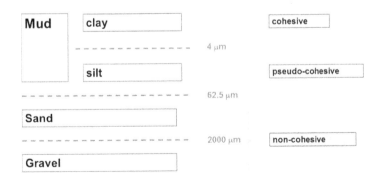

FIGURE 2.1
Sediment size schematic. (After Van Maren 2009.)

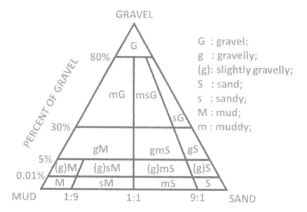

FIGURE 2.2
Sediment textural groups triangle. (After Folk 1954.)

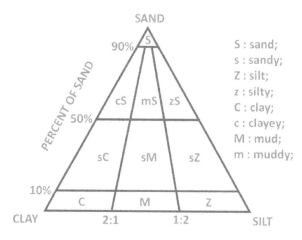

FIGURE 2.3
Sand–silt–clay textural groups triangle. (After Folk 1954.)

A detailed classification of sediments with size larger than the nominal size for gravel would require perhaps a third gravel–cobble–boulder triangle for seabeds with a large content of gravel, cobbles, and boulders. Such a classification could be relevant in very energetic regions with a cobbled seabed. Some of these environments are found in high latitudes, for example, in the Orkney Islands, UK, or in rocky-cliffed regions where a large number of boulders and cobbles may detach from cliffs into the sea. However, sediment entrainment occurs at speeds that are not large enough, in general, to entrain boulders into the flow, and therefore we will not elaborate more into this idea of a gravel–cobble–boulder textural group triangle.

Sediment entrainment depends on the shear stress, τ, as introduced in Chapter 1. The dimensionless shear stress, τ^*, also known as the Shields parameter, θ, may be expressed as

$$\tau^* = \theta = \frac{\tau}{(\rho_s - \rho)gD},\qquad(2.1)$$

where ρ_s is the density of the sediment; ρ the density of the fluid; g the acceleration due to gravity; and D the characteristic diameter of the particle, usually the median grain diameter, d_{50}. Physically, θ is proportional to the ratio of (1) the force applied by the fluid on the particle and (2) the weight of the particle. When a particle is resting on the seabed, additional particle-to-particle contact forces maintain the sediment in an equilibrium position in relation to the bed. Closest to the particle surface, viscous shear stresses act on the boundary, and the boundary layer flow field $U(z)$ follows, approximately, a viscous linear profile:

$$U(z) = \frac{u_*}{\delta_\nu}z,\qquad(2.2)$$

where $\delta_\nu = \nu/u_*$ is the viscous length and u_* the friction velocity. Further away from the surface, the fluid forces produce a positive (upstream) and negative (downstream) pressure field around the particle. The sum of the viscous and pressure forces results in the total fluid force acting on the particle (Southard 2006). This total fluid force may also be separated into drag (parallel to the bed) and lift (perpendicular to the bed) forces. The viscous sublayer is thinned and accelerated when high-speed flows impinge on the particle, and when the turbulent lifting forces are large enough, when $\theta > \theta_c$, the critical Shields parameter, the sediment is entrained into the flow.

From μ, the fluid viscosity, and the variables ρ, ρ_s, τ, D, and $(\rho_s - \rho)g$, two dimensionless parameters may be constructed. The first has already been discussed: the Shields parameter. The second is obvious: the ratio ρ_s/ρ. A third one involving these variables and the friction velocity $u_* = \sqrt{\tau_0/\rho}$, τ_0 being the bottom shear stress, was only mentioned in passing in Chapter 1: the boundary Reynolds number,

$$Re^* = \frac{\rho u_* D}{\mu}.\qquad(2.3)$$

The condition for incipient sediment motion may be expressed as a surface in a 3D space involving θ_c, Re^*, and ρ_s/ρ. But for constant ρ_s/ρ, this surface reduces to

$$\theta_c = f(Re^*).\qquad(2.4)$$

A number of experiments have been performed to characterise the dimensionless critical shear stress in Equation 2.4 for various Reynolds number regimes. Figure 2.4 synthesises 100 years of research (Buffington & Montgomery 1997), on dimensionless shear stress observations at the incipient threshold of motion, based on two different techniques, the watch-the-bed method (black lines), and the reference-transport-rate method (grey lines).

The names of the methods are self-explanatory. The watch-the-bed method consists of watching a plane bed subject to uniform flow with constant speed, increasing this flow speed, and identifying the moment when sediment motion starts to occur. As Southard (2006) highlights, this is a very subjective approach, which may explain the scatter in the threshold data reported when using this method. One of the problems is that there are a number of situations where there is weak but noticeable bed movement, for example, when a chaotic eddy hits the bed. The reference-transport-rate method is an indirect approach, based on measurements of the sediment transport rate. This method was used, for example, by Shields (1936), in his pioneering work on the threshold of motion for gravelly river beds. The reference-transport-rate method consists of making several experiments at different flow strengths above the threshold of motion, measuring the dimensionless unit transport rate in all of these experiments and, once a curve has been obtained, finding (by interpolating or extrapolating the bed shear stress measurements) the value of the bed shear stress associated with the reference transport rate (Southard 2006). The flow strength, regarding the Shields parameter, θ, and the seabed roughness, k_s, are linked to the mobility number, which is in turn defined as (Van Rijn 2007, Davies & Robins 2017)

$$\psi = \frac{U_c^2 + U_w^2}{(1-s)gd}, \tag{2.5}$$

where U_c is the current speed, U_w is the wave speed, $s = \rho_s/\rho$ is the relative sediment density (with ρ_s the sediment density and ρ the flow density), and $d = d_{50}$ is the median grain diameter.

Figure 2.4 shows there are three Reynolds number regimes. The first consists of a smooth turbulent flow regime, for $Re_c^* < 2$, where the dimensionless critical Shields parameter for the median grain size, $\theta_c = \tau_{c50}^* = \tau_{c50}/(\rho_s - \rho)gD_{50}$, decays as a power law with increasing boundary Reynolds

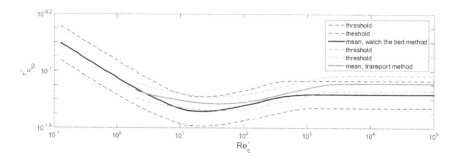

FIGURE 2.4

$\tau_{c50}^* = \theta_c$ vs. Re_c^*, using the reference-transport-rate method (starting from Re_c^* above 1) and the watch-the-bed method (starting from $Re_c^* = 0.1$). (After Buffington & Montgomery 1997.)

number $Re_c^* = u_c^* k_s / \nu$; ρ_s is the sediment density, ρ the water density, g the gravitational acceleration, D_{50} the median grain diameter of the sediment, u_c^* the critical shear velocity for incipient motion, k_s the boundary roughness length scale, and ν the kinematic viscosity. As $Re_c^* \to 0$, θ_c tends to a constant value lying between 0.16 and 0.45, which is consistent with the value reported by Soulsby (1997). Then, one may observe a transitional regime, which occurs in the range $2 < Re_c^* < 2{,}000$ in both the reference-transport-rate method and the watch-the-bed method. Finally, in the rough turbulent regime, when $2{,}000 < Re_c^*$, the critical Shields stress plateaus to a constant value. However, there is a lot of scatter in the data, and there is a difference between the two methods, with θ_c larger in the reference-transport-rate method, where $\theta_c = 0.069 \pm 0.017$, than in the watch-the-bed method, where $\theta_c = 0.0515 \pm 0.0215$. Therefore, θ_c may vary between 0.03 and 0.086, which, as Buffington & Montgomery (1997) highlight, calls for caution when defining a threshold of motion.

Other researchers have used excess pressure gradients to analyse the threshold of motion. This is based on observations under energetic wave conditions, when the mobile bed layer becomes thicker, and the threshold of motion does not depend any more on the diameter of the sediments at the bed but on the horizontal pressures applied on this mobile layer. In other words, bed dilation occurs when the horizontal pressure gradient dominates over the sediment's immersed weight. Sleath (1999) defined the threshold of motion under such conditions through the Sleath parameter,

$$S_\omega = \frac{\rho U_o \omega}{(\rho_s - \rho) g}, \tag{2.6}$$

with U_o and ω being the wave velocity amplitude and the angular wave frequency, respectively, and

$$\frac{\partial p(t)}{\partial x} = \rho U_o \omega \sin(wt)$$

is the sinusoidal wave horizontal pressure gradient at time t, in the x-direction. Thus, the Sleath parameter is 'the ratio between the destabilizing force applied by the peak horizontal pressure gradient to the stabilizing force applied by gravity' (Foster et al. 2006), and has no dependence on the grain size. The Sleath parameter, however, may be generalised to an instantaneous Sleath parameter that depends explicitly on the horizontal pressure gradient (Foster et al. 2006), as

$$S_\omega(t) = \frac{-1}{(\rho_s - \rho) g} \frac{\partial p(t)}{\partial x}. \tag{2.7}$$

In shallow wave environments,

$$-\frac{\partial p}{\partial x} = \rho \frac{\partial u_\infty}{\partial t} + \rho u_\infty \frac{\partial u_\infty}{\partial x},$$

or the material derivative of the free-stream velocity, u_∞. In the limit of linear wave theory, the advective term may be neglected. Negative magnitudes of S imply the strengthening of the onshore-directed flow, while positive magnitudes imply the strengthening of the offshore-directed flow. Thus, the horizontal pressure gradient has a significant influence on the incipient motion of sediments in the surf zone; this gradient is linked to wave-orbital velocities from pitched-forward waves that reach the sea floor. So when the waves have these characteristics, the Sleath parameter becomes more important than the Shields parameter as a measure of the sediment threshold motion. However, there are circumstances when both pressure gradients and shear stress forces acting on the bed may be responsible for sediment mobilization, for example, in the case of coarse sediments being mobilized by regular, shoaling waves (Terrile et al. 2006).

The shear stress at the bed is difficult to predict. In depth-averaged tidal models for the continental shelf, it is generally taken to depend on a dimensionless drag coefficient C_D. In some models, it is set to a constant, as default. Generally, $C_D = 0.003$ in water (Roelvink et al. 2015), and $C_D = 0.0017$ in air (Kullenberg 1976). C_D in air is the value at 10 m above ground. However, there are different formulations for the drag coefficient: the Chézy, the Manning, and the White–Colebrook grain size formulations. Most coastal regions, at local or regional scales, have variable sediment composition; therefore the bottom roughness ideally should also be spatially varying. However, this is not always practical, and a certain homogeneity is assumed in most cases.

The drag coefficient, like the bottom roughness, depends on the height and geometry of bottom bedforms such as ripples and seabed mounds. The drag coefficient and bottom roughness at Duck, North Carolina, are both larger in the surf zone than seawards of the surf zone (Feddersen et al. 2003). Moreover, the drag coefficient depends on bottom roughness as we move seawards from the surf zone, but it seems the two of them are uncorrected within the surf zone. In flows with horizontal velocity variations in the vertical direction, the shear stress in the bottom boundary layer (BBL) is best defined in terms of the vertical diffusivity, K_M, and the velocity gradient:

$$\tau_b = K_M \frac{\partial u}{\partial z} = \rho u_*^2, \tag{2.8}$$

with u_* the friction velocity, as seen at the beginning of Section 2.2.1. Some models may use the simple drag coefficient formulation to determine the shear stresses, while others consider complex wave and current interaction processes over a movable bed and are based on the more general Equation 2.8. Some models have the ability to use both. Either method may be suitable, according to the purpose of the modelling effort. Either way, for flat rough beds, the seabed type alone defines the bed roughness regime of the flow, which affects significantly the current velocity fields close to the bed.

So far we have highlighted two mechanisms which are crucial for sediment transport in the BBL, namely the bottom shear stresses and the pressure gradients. Turbulence, diffusion, and drag also play an important role, with diffusion and drag being linked to the bottom shear stress, as seen in Equation 2.8. Diffusion has a mechanical and a turbulent component, and these two components are statistically independent from one another, so they can be added to obtain the total diffusion (Nepf 1999). We now introduce some fundamental concepts regarding turbulence in coastal environments, based on the notes by Westerink (2003). First, the flow velocity components, the pressure, and the density may be divided into a mean turbulent component and a turbulent fluctuation, namely:

$$\alpha = \bar{\alpha} + \alpha', \tag{2.9}$$

where α refers to any of the three velocity components (u, v, w), to density ρ, or to pressure p, and the time-averaged term,

$$\bar{\alpha} = \frac{1}{T} \int_0^T \alpha \, dt, \tag{2.10}$$

is averaged over a time period, T, which is large compared to the turbulent fluctuations but small compared to the timescale of variation of the turbulent time-averaged flow. The flow is assumed to be ergodic over T; that is, the flow statistics are stationary over that time period. Also, the turbulent fluctuations dissipate and mechanical energy turns into heat when the Reynolds number is of order 1. This is also known as the *Kolmogorov dissipation scale*.

The details of the computations leading to the conservation equations are described in Westerink (2003); here we will simply state the main results. The mass conservation equation may be split into two equations, one involving the mean velocity components and another one involving the turbulent fluctuations:

$$\frac{\partial \bar{u}}{\partial x} + \frac{\partial \bar{v}}{\partial y} + \frac{\partial \bar{w}}{\partial z} = 0 \text{ and} \tag{2.11}$$

$$\frac{\partial u'}{\partial x} + \frac{\partial v'}{\partial y} + \frac{\partial w'}{\partial z} = 0, \tag{2.12}$$

respectively. In the case of the momentum conservation equations, a balance is obtained which involves inertial, viscous, and fluctuation terms, in each of the three spatial directions, which in vector notation are expressed as

$$\frac{D\bar{U}}{Dt} = \frac{\bar{\rho} g \hat{k}}{\rho_0} - \frac{1}{\rho_0} \nabla \bar{p} + \frac{1}{\rho_0} \nabla \cdot \mathbf{T}_{t/m}, \tag{2.13}$$

where D/Dt is the material derivative, including the time gradient and the advection terms; $\bar{U} = \bar{u}\hat{i} + \bar{v}\hat{j} + \bar{w}\hat{k}$ is the turbulent time-averaged velocity vector; $\dfrac{\bar{\rho} g \hat{k}}{\rho_0}$ is a compressibility term; $-\dfrac{1}{\rho_0} \nabla \bar{p}$ is the pressure gradient term;

$\nabla = \dfrac{\partial}{\partial x}\hat{i} + \dfrac{\partial}{\partial y}\hat{j} + \dfrac{\partial}{\partial z}\hat{k}$ is the gradient vector; and $\dfrac{1}{\rho_0}\nabla \cdot \mathbf{T}_{t/m}$ is the turbulent/molecular (t/m) tensor. The t/m tensor is of the form

$$\mathbf{T}_{t/m} = \tau_{i,j}^{t/m} = \rho \left[\nu \frac{\partial u_i}{\partial x_j} - \overline{u_j'u_i'} \right], \qquad (2.14)$$

where

- $\nu\dfrac{\partial u_i}{\partial x_j}$, the *viscous stresses*, represent the averaged molecular motions. These terms are necessary in models that do not simulate momentum transfer through direct molecular collisions.

- $-\overline{u_j'u_i'}$, the *turbulent Reynolds stresses*, represent the averaged effect of momentum transfer due to turbulent fluctuations. These terms are necessary in models based on turbulent time-averaged variables.

Sand transport by currents is strongly dependent on the turbulence variances

$$\overline{u'u'}, \overline{v'v'}, \text{ and } \overline{w'w'},$$

which are the diagonal elements of the Reynolds stress tensor. The total turbulent kinetic energy is given by

$$E = \frac{1}{2}\rho \left(\overline{u'^2} + \overline{v'^2} + \overline{w'^2} \right). \qquad (2.15)$$

The other three components of the Reynolds stress tensor, namely

$$-\rho\overline{u'w'}, -\rho\overline{u'v'}, \text{ and } -\rho\overline{v'w'},$$

are each dominated by different processes. For example, $\overline{u'w'}$ is dominated by sweeps ($u' > 0$, $w' < 0$) and bursts ($u' < 0$, $w' > 0$). Sweeps and bursts are event-based processes which may initiate sediment resuspension (Salim et al. 2017). Understanding the contributions of sweep and burst events to sediment transport in the bottom boundary layer is critical, but very few researchers have studied them in detail. The work by Salim et al. (2017) was based on acoustic Doppler Velocimeter (ADV) experiments. The ADV was used to analyse sediment resuspension processes, through instantaneous velocity and acoustic backscatter measurements. They concluded that sweeps and bursts contributed more to total sediment flux than acceleration or deceleration events. The role of turbulence in sediment transport was first studied by Nelson et al. (1995), who also worked on this problem using experimental techniques. Nelson et al. (1995) analysed the effects of sweeps and bursts on the bottom shear stress. According to their study, sweeps have a positive effect on the bed shear stress and contribute significantly to the resuspension of sediment, mostly because they are very common. Nelson et al. (1995) define two more processes: outward interactions ($u' > 0$, $w' > 0$) and inward interactions ($u' < 0$, $w' < 0$). They found that outward interactions are relatively rare, and although they contribute negatively to the bed shear stress, individually they move as much sediment as sweeps of comparable magnitude and duration. Bursts and inward interactions are the processes that contribute the least to sediment transport. The observations of Nelson et al. (1995) and Salim et al. (2017) both imply that bed shear stress formulations work well in well-developed boundary layers, but turbulence is an important process during flow separation, which is in turn important during the formation of bedforms.

2.2 Sediment Transport

Once the threshold of motion is exceeded, the lifting forces on the sediment may carry sediment into suspension, and small bedforms, called *ripples*, start to form. In the rippled bed regime, when $\theta_c < \theta < 0.8$, suspended sediment concentrations (SSCs) in the bed boundary layer can be parameterised in terms of the excess shear stress (Foster et al. 2006), while for $\theta \sim 0.8-1.0$, the ripples disappear, and a mobile, sheet-flow layer forms. As predicted by Sumer et al. (1996) and observed by Dohmen-Janssen & Hanes (2002, 2005), the sheet-flow layer thickness increases linearly with θ. For skewed and asymmetric waves, this thickness may vary from 10 to 20–40 grain diameters (Hsu & Hanes 2004), depending on the wave conditions. It is important to note that erosion depth and sheet-flow layer thickness are both larger for fine sand than for coarse sand (Dohmen-Janssen et al. 2001), maybe contrary to intuition.

2.2.1 Flow Over a Flat Bottom

As defined in the previous sections, the friction velocity at the bed is $u_* = \sqrt{\tau_b/\rho}$, which is proportional to the square root of the bottom shear stress when the density is constant. However, further away from the bed, the velocity in a turbulent flow depends as well on the height above the bed, z, following a typical logarithmic profile, namely the *law of the wall* (Raupach et al. 1991):

$$U(z) = \frac{u_*}{\kappa} \ln\left(\frac{z}{z_0}\right). \tag{2.16}$$

Here $\kappa \sim 0.4$ is the Von Kármán constant, and z_0 is a reference height of zero mean velocity, called the *hydrodynamic roughness*, and found by extrapolating the log-velocity profile to the bed. When the characteristic grain diameter, D, is small enough, the very near bed boundary layer comprises a viscous sublayer (Charru et al. 2013), with a velocity profile that grows linearly with height above the bed:

$$U(z) = \frac{u_*}{\delta_\nu} z = \frac{u_*^2}{\nu} z, \qquad (2.17)$$

where $\delta_\nu = \nu/u_*$ is the viscous length. Under these conditions, the flow is hydrodynamically smooth, the boundary layer thickness is $\sim 6\delta_\nu$, and the hydrodynamic roughness is $z_0 \sim 0.11\delta_\nu$.

For large grain diameters, say, $D \sim 10\delta_\nu$, the flow becomes hydrodynamically rough, and the hydrodynamic roughness becomes dependent on the grain diameter, as $z_0 \sim 0.03 - 0.1D$ (Andreotti 2004) or, typically, $z_0 \sim D/12$. However, if we have a moving sediment layer at the surface, then z_0 may be larger than this.

2.2.2   Flow Over a Wavy Bottom

The simplest wavy bottom is a 2D sinusoidal wave with small amplitude $\zeta = z_0 \cos(kx)$. Over a wavy bottom, as the hydrodynamic roughness and the free stream velocity increase, one may observe four different flow regimes, schematised in Figure 2.5: a viscous-flow regime throughout the domain; a laminar–inertial flow regime, with a viscous-flow layer near the bed and an inertial-flow layer at larger heights; a transitional regime, with a turbulent-flow layer developing between the viscous-flow and the inertial-flow layers; and a turbulent regime, where a very thin viscous-flow layer near the bed surface is followed by a thick, turbulent-flow layer and then finally an inertial-flow layer further away from the bed. An inertial flow is a flow that, without any applied external forcing, continues in motion due to the momentum it acquired previously. These four flow regimes are governed by the magnitude of the boundary Reynolds number, as explained above. However, as Charru et al. 2013 note, they can be characterised as well in terms of the product of kz_0, with k the wave number and z_0 the hydrodynamical roughness. $kz_0 \ll 1$ corresponds to the long wave limit, also characterising the turbulent flow regime; the transitional regime, between laminar and turbulent flow, is characterised by $10^{-5} < kz_0 < 10^{-3}$.

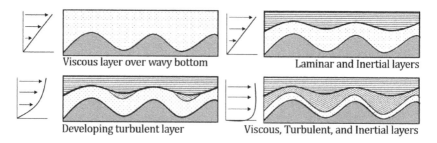

Viscous layer over wavy bottom   Laminar and Inertial layers

Developing turbulent layer   Viscous, Turbulent, and Inertial layers

FIGURE 2.5
Schematic of flow regimes over a wavy bottom.

Depending on the slope kz_0 of the sinusoidal bottom, the flow may be linear or nonlinear, with $0.3 < kz_0$ corresponding to the case of flow separation (Charru et al. 2013). Flow separation is the most complex case from a theoretical point of view, because of the development of complex turbulent structures and the formation of recirculation bubbles on the lee side (defined in Figure 1.2 of Chapter 1) of the wavy bottom. The Kelvin–Helmholtz instability induces the formation of spanwise vortices that impinge on the bottom, near the reattachment point. This dissipates the energy from the waves in the BBL. On erodible beds, sediment mobility deforms the topography, which becomes asymmetric. Mature ripples and subaqueous dunes deform until the height-to-length ratio reaches an equilibrium value of about $1/15$. Over wavy bottoms, the bed roughness scale is much larger than over flat beds. For example, for rippled sand beds, $z_0 = 0.6$ cm, as opposed to $z_0 = 0.04$ cm. This is due to the bedform *form drag*, i.e., the horizontal force due to the adverse pressure gradient on the lee side of the bedform, with skin friction operating mostly on the stoss side (also defined in Figure 1.2 of Chapter 1). To complicate matters even more, the bed roughness also depends on the lee side slope, with low roughness for slopes $< 10°$ and high roughness for slopes larger than $20°$ (Lefebvre & Winter 2016).

2.3 Hydrodynamics in the Coastal Zone

2.3.1 Structure of the Marine Boundary Layer

The traditional view of the BBL is depicted in Figure 2.6, which is accurate from a climatological point of view; that is, on average, over a long time, a logarithmic velocity profile just above the bed layer is a relatively good approximation, particularly under steady current conditions. However, it has been shown that long, shallow-water waves with a daily cycle, such as seiches in lakes, can change significantly the near-bottom current profiles and the turbulent dissipation rate dynamics (Lorke et al. 2002). An alternative to the simple power law profiles common in the literature is to use a boundary layer profile for the current or wind horizontal speeds, $U(z)$, that depends on three constants, locally adjusted to get the closest fit to the data (Gross & Magar 2015), namely

$$U(z) = U(10) \left(\frac{z}{10} + cz \right)^{a + b/z} . \tag{2.18}$$

This provides a better fit to the vertical profiles of the horizontal speeds than the traditional power law

$$U(z) = U(10) \left(\frac{z}{10} \right)^{\alpha} ,$$

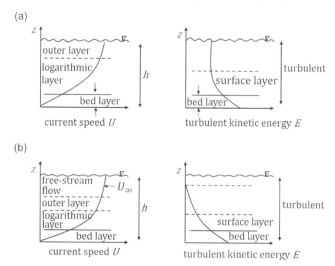

FIGURE 2.6
Structure of the marine boundary layer. (a) Depth-limited flow and (b) unconfined depth flow.

particularly for tidal or wind energy applications. Indeed, generally neither marine nor wind turbines are at 10 m heights above the sea floor or the sea surface, respectively.

2.3.2   Open Coast Dynamics

On open coasts or in very large lakes, the dynamics are driven by nearshore currents, tides, swell waves, and storm waves. When nearshore processes are predominantly cross-shore, they are expected to be mainly wave driven. Thus, we will summarize first nearshore wave dynamics and then follow with some considerations on nearshore currents.

2.3.2.1   Stochastic and Deterministic Representation of Wave Fields

Waves are generated by shear stresses and pressure effects applied by the wind on the surface of the ocean. Waves may be divided into wind waves and swell waves, depending on whether they are locally generated or whether they have travelled a distance between the generation area and the observation area. Wind and swell waves have different characteristics; for example, wave height time series for wind waves at the point of generation show a signal that is distinctly random, while swell waves are waves with same wavelength and wave heights. Figure 2.7 shows typical waves based on their periods. Their generating and restoring forces are also shown, and this is crucial information

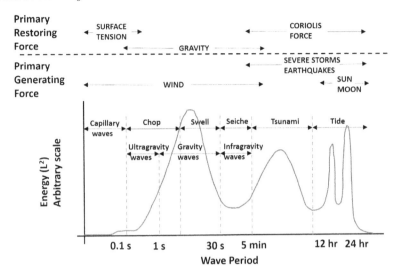

FIGURE 2.7

Different types of ocean waves according to their characteristic period but, most importantly, according to their primary generating and primary restoring forces.

in the analysis of different types of waves. For example, chop and swell are wind waves which are little affected by surface tension or Coriolis forcings. But they are similar in that they are generated by the wind, and wind speed and direction are random by nature, so wind waves have random amplitudes and periods in the region where they are generated. In contrast, capillary waves, which are waves with small periods and amplitudes, are affected by surface tension; tides are generated by the forces exerted by the Sun and the Moon over the ocean, which are fully predictable, but their propagation is affected by the continents and by friction with the seafloor in shallower areas.

The fact that we can depict waves in a single 'name vs. range of periods' diagram can be misleading, because the generating and restoring forces, as well as the evolution of these waves as they propagate across the ocean, are distinctive to each type. This will be discussed in detail, but before that, it is useful to define parameters and variables commonly used in wave dynamics. These parameters are shown in Figure 2.8. From four basic parameters: wave period T, wave height H, wavelength L, and water depth h, one can define the wave phase speed, $c = L/T$; the wave steepness, H/L; the wave number, $k = 2\pi/L$; and the wave frequency, $\omega = 2\pi/T$. Waves are generated during storms, by winds blowing on the surface of the sea. These are called *wind waves* and are random in nature, with nonunique wave height and period. As waves start travelling at different phase speeds c, they disperse into swell, with one distinct H and L, similar to the wave shown in Figure 2.8. The only wave parameter that remains constant for an individual wave train as it propagates

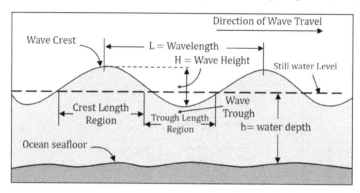

FIGURE 2.8
Parameter definitions for wave analysis.

across the ocean into shallower regions is the wave period (and obviously the wave frequency, as it is the inverse of the period). These parameters can be further combined into what is known as the *wave dispersion equation*, with the general form

$$c = \frac{gT}{2\pi} \tanh\left(\frac{2\pi h}{L}\right), \qquad (2.19)$$

which can be simplified to

$$c_o = \frac{gT}{2\pi} \qquad (2.20)$$

in deep water and to

$$c = \sqrt{gh} \qquad (2.21)$$

in shallow water.

Waves locally generated by wind depend on fetch, local water depth, wind speed and direction, and tidal currents and water level changes, among other factors. In deep water, waves are only weakly nonlinear, so although waves are generally random, with short, medium, and long period waves superimposed at and near the region of generation, they can be described as the sum of a large number of independent components, which can be represented as a probability density distribution. Only when individual waves have separated through dispersion into swell can they be described with a single wave height, period, and direction. The process of dispersion is illustrated in Figure 2.9. Wave heights at given observation points can be characterised in time domain or frequency domain; models using time domain are phase-resolving models, while frequency-domain models are phase averaged. That is, the individual wave components are not known, in principle. Representative wave conditions can be obtained using statistical tools, both in deterministic (time domain) and spectral (frequency domain) wave modelling. In the frequency-domain framework, waves are described as a *wave spectrum*, with wave energy used

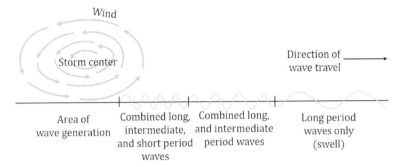

FIGURE 2.9
Progression from wind waves on the right to swell on the left.

as a proxy for wave height. The representative height in random seas is H_s, the *significant wave height*, which corresponds to the highest one-third of the waves,

$$H_s = 4(m_0)^{1/2},\qquad(2.22)$$

where m_0, the variance of the wave elevation (Cuadra et al. 2016), is the zeroth order spectral moment of the spectral density function $S(f)$, or *waves spectrum* (Ochi 1998). The nth order moment, m_n, is defined as

$$m_n = \int_0^\infty f^n S(f) df.\qquad(2.23)$$

Wave heights may be defined as the *zero-up-cross* wave height or as the height of the *crest-to-trough* waves (Goda 1970). In the former case, the wave period is also the period between zero-up-crossings, while in the latter, the wave period is the time elapsed between two wave maxima. These two definitions are illustrated in Figure 2.10.

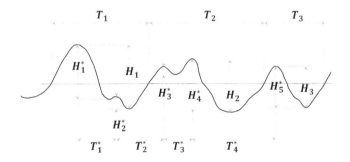

FIGURE 2.10
Wave heights and periods for crest-to-trough waves (H_i^* and T_i^*) and zero-up-cross waves (H_i and T_i).

Some mean wave periods are defined in terms of the moments m_n, in the form

$$T_{x,y} = T_{mxy} = \frac{m_x}{m_y}, \tag{2.24}$$

such as the wave energy period,

$$T_e = T_{-1,0} = T_{m-10} = \frac{m_{-1}}{m_0}; \tag{2.25}$$

an estimate of the mean period used in the design of marine structures,

$$T_{0,2} = T_{m02}; \tag{2.26}$$

or an estimate of the mean period used in freak waves and wave forecast research,

$$T_{0,1} = T_{m01}. \tag{2.27}$$

Other wave periods are also of importance, for example, the peak period,

$$T_p = \frac{1}{f_p}, \tag{2.28}$$

with f_p the spectral peak frequency, when $S(f)$ is maximum, or the average zero-crossing period,

$$T_z = \sqrt{\frac{m_0}{m_2}}. \tag{2.29}$$

If the period and the significant wave height can be written in terms of m_n, then so can be other quantities that depend on them. For example, the wave energy flux or power density per meter of wave crest,

$$P = \frac{\rho g^2}{4\pi} m_{-1} = \frac{\rho g^2}{4\pi} H_s^2 T_e, \tag{2.30}$$

can either be determined using H_s and T_e, or m_{-1}.

There are two well-known, preferred wave spectrum shapes, the Pierson–Moskowitz (PM) spectrum for fully developed seas and the JONSWAP (Joint North Sea Wave Project) spectrum for developing seas (Hasselmann et al. 1973). For both of them, the wave energy density $S(\sigma)$, is expressed in terms of the intrinsic angular frequency, $\sigma = 2\pi/f$,

$$S(\sigma) = \alpha g^2 (2\pi)^{-4} \sigma^{-5} \exp\left[-\frac{5}{4}\left(\frac{\sigma}{\sigma_p}\right)^{-4}\right] \gamma_0^{\exp\left[-\frac{1}{2}\left(\frac{\sigma-\sigma_p}{\epsilon\sigma_p}\right)\right]}, \tag{2.31}$$

where α is an energy scaling factor (0.0081 for PM); σ_p is the peak angular frequency; γ_0 is a peak enhancement factor (1 for PM and 3.3 for JONSWAP); and ϵ determines the spectral width around the peak (0.07 for $\sigma < \sigma_p$ and

0.09 for $\sigma > \sigma_p$). The spectral peakedness, Q_p, of the wave spectrum, was defined by Goda (1970) as

$$Q_p = \frac{2}{m_0^2} \int_0^\infty f S^2(f) df, \qquad (2.32)$$

a dimensionless quantity independent of time. When the spectrum has a single peak at $f = f_p = f_0$, the spectral peakedness is approximately given by

$$Q_P = \frac{2f_0}{f_4 - f_3}, \qquad (2.33)$$

where f_3 and f_4 satisfy $S(f_3) = S(f_4) = S(f_0)/5$ (Goda 1970). The JONSWAP spectrum is for developing seas and, as such, has a number of characteristics: it is a fetch-limited spectrum, the fetch being the distance between the location of computation and the land; the JONSWAP spectrum can evolve indefinitely due to nonlinear wave–wave interactions (Hasselmann 1966). So waves may grow with distance or time, depending of the values of the parameters in the JONSWAP spectrum.

A frequency directional spectrum may be constructed as

$$S(\sigma, \theta) = S(\sigma) D(\theta, \theta_p), \qquad (2.34)$$

with $D(\theta, \theta_p)$ a directional distribution, or *directional energy spreading function*, dependent on the wave direction, θ, and the peak direction, θ_p. Wave direction is the direction the waves come from, just like with wind (the main wave generating force); it could be measured with respect to the geographic north, the magnetic north, or the incident wave angle with respect to the normal to the shore. A number of theoretical expressions for $D(\theta, \theta_p)$ have been proposed by a number of authors, and several techniques for evaluating the function from measured buoy data have also been proposed. When *maximum likelihood* (ML) methods are used to estimate D from measurements, they seem to give excellent results (Ochi 1998), when compared with estimates obtained from Fourier series expansions.

For deterministic wave frameworks, the surface elevation can be generated using a random phase model where a large number of sinusoidal wave components with amplitude η_j are calculated from the frequency spectrum (Equation 2.31), and assigned a random phase Φ_j. The resulting surface elevation is of the form

$$\eta(x, y, t) = Re \sum_{j=1}^N \eta_j \exp^{i(\sigma_j t - k_{x,j} x - k_{y,j} y + \Phi_j)}, \qquad (2.35)$$

which, at the point of generation, has a 'random seas' or 'wind waves' signature and visually looks like a random signal, similar to that illustrated in Figure 2.11. This particular wave data was measured on 24 November 1987

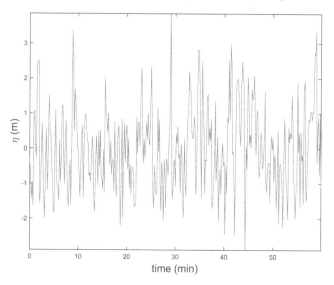

FIGURE 2.11
Example time series of an offshore random sea state, measured with an ultrasonic wave gauge with 1 Hz sampling frequency (plotted every 10 s).

at the Poseidon Platform (see http://mararchief.tudelft.nl/file/51379/), 3 km off the Yura coast, in the Japan Sea.

Time domain representations are phase resolving and are relevant for local modelling at short timescales. Such descriptions are common in volume-of-fluid, direct numerical simulations, some Boussinesq and some nonhydrostatic models. However, for sediment transport and morphodynamics at regional scales, usually at lengthscales of kilometres and timescales of the order of a week, phase-resolving methods are in most cases too expensive from a computational standpoint, and therefore one resorts to frequency-domain representations and wave action balance equations to analyse the evolution of the surface waves field. In such models, the intrawave scales are not resolved explicitly any longer. This means that it may be necessary to introduce parameterisations in the model to capture the effects of wave asymmetry, wave skewness, wave breaking, turbulence, and run-up, for example, on sediment transport.

Nonlinear processes in wave dynamics are fundamental in risk analysis; extreme events in the ocean (e.g., freak waves) may be driven by wave nonlinearities, and these wave nonlinearities are more pronounced during extreme events; other extreme events are driven by atmosphere–ocean interaction during storms, i.e., during extreme events in the atmosphere. Instabilities in the ocean are generated by different mechanisms, including waves encountering an opposing current, nonlinear wave–wave interactions, linear focusing, wave interactions with the bathymetry, and the fetch or fetch length, amongst others. In the next section, we will discuss some of these phenomena in more detail.

2.3.2.2 Wave Transformation Processes Caused by Obstacles and Bottom Topography

Seven common wave transformation processes occur in the surf zones of the ocean nearshore (they may also be observed in deep water, but to a lesser extent, and in very specific cases): diffraction, refraction, reflection, shoaling, wave breaking, and wave set-up and run-up. During these wave transformation processes, either the wave speed, the wave direction, or the wave height changes over space.

Diffraction is the process of radiation of wave energy when a wave encounters an obstacle or an opening between obstacles, causing wave bending. The obstacles may include engineering structures, such as breakwaters or harbour entrances, or natural structures such as islands or inlets. It is important to note that when a wave train encounters a current with a speed equal or greater than the wave group speed, the current acts as a virtual obstacle to wave propagation, causing wave diffraction and breaking (Reeve et al. 2018). Such obstacles separate an incoming wave zone from a shadow zone, with wave diffraction calculations assessing the wave energy that is 'leaked' into the shadow zone (Kamphuis 2010). A wave diffraction and reflection diagram is shown in Figure 2.12. The diagram shows three distinct regions: the incident waves or incidence region (regions 1 and 2), including the region of incoming waves and the region undisturbed by the presence of the obstacle; the shadow or diffraction region (region 3), behind the obstacle, where the waves are diffracted; and the reflection region with short-crested waves (region 4),

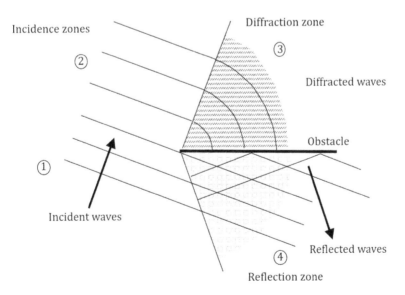

FIGURE 2.12
Wave diffraction and reflection diagram. (Based on Reeve et al. 2018.)

where incident and reflected waves interact with each other to form a sea surface pattern known as *clapotis gaufre*, associated with superposition of oblique incident waves and their reflection. Although such computations may be complicated, one may use diffraction templates, based on linear wave theory with no refraction (constant water depth), to evaluate the diffraction coefficient, K_d (Goda 2010). Mathematical formulations of the diffraction problem may also be formulated, adapted to the geometry of the obstacle interfering with the propagation of the waves. For example, Zhu & Mitchell (2009) analyse the process of wave diffraction around and inside a floating hollow cylindrical structure using potential flow theory, but there are very few studies analysing the effects of wave diffraction on sediment transport and morphology. In fact, nonlinear spectral wave models are very poor at capturing the diffraction process, and wave-resolving methods frequently analyse the sediment transport and morphodynamics in scenarios when diffraction is not separated from the refraction process, for example, wave behaviour around headlands and their impacts on nearby beaches or morphological studies of embayed beaches (a beach between two headlands) as in (Mascagni et al. 2018).

Refraction is the process of wave bending due to interaction with the seabed. As waves propagate nearshore and reach regions shallower than the wave base, corresponding to a water depth that is half the wavelength, the waves begin to slow down, and the wave crest starts to align itself with the bathymetric contours. The refraction coefficient, K_R, may be computed by applying Snell's law to ocean waves approaching a medium with changing depth. K_R is defined as

$$K_R = \frac{b}{b_0} = \frac{\cos\alpha}{\cos\alpha_0}. \tag{2.36}$$

Figure 2.13 shows the process of wave crest alignment towards the shore, due to wave refraction caused by the interaction of the waves and the bathymetry. Wave refraction causes waves to arrive perpendicular to the coast as they align with the bathymetry. At headlands and embayed beaches, wave refraction causes wave convergence at the headland and wave divergence at the beach, resulting in high-energy conditions and wave-driven erosion near the headland, and sediment deposition at the beach. So with enough time (and assuming no human intervention), wave action will tend to create a smooth coastline, unless the headland is rocky and difficult to erode, in which case sediments will get deposited at the headland base, as observed in beaches protected on both ends by rocky groynes—as shown, for example, in Figure 2.14, a sandy beach protected by two rock groyne breakwaters on East Gibraltar.

As waves refract, they also become steeper, a process known as *wave shoaling*. Waves shoal due to conservation of wave energy flux, which causes the wave height to increase as the waves propagate into shallower water. If H_i is the wave height at depth h_i, with h_i decreasing as i increases, then

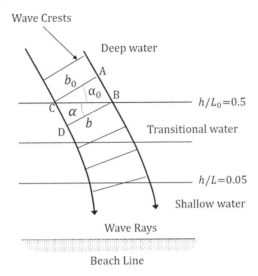

FIGURE 2.13
Wave refraction diagram. (Based on Reeve et al. 2018.)

FIGURE 2.14
Nicely shaped Sandy Beach with two peninsulas on east Gibraltar. View from top.

$$\frac{H_{i+1}}{H_i} = \left(\frac{\cos \alpha_i}{\cos \alpha_{i+1}} \right)^{1/2} \left(\frac{c_{gi}}{c_{g(i+1)}} \right)^{1/2}, \tag{2.37}$$

where c_{gi} is the group speed and α_i is the angle between the wave crest and the isobath at depth h_i. The shoaling coefficient, K_S, is defined as

$$K_S = \left(\frac{c_{gi}}{c_{g(i+1)}} \right)^{1/2}. \tag{2.38}$$

Refraction and shoaling of waves cause alterations of the wave height, wavelength, and shape and ultimately wave breaking near the shoreline. Once waves have broken, their energy is transferred to the ocean in the form of cross-shore and longshore currents and a water-level set-up, and the interactions of the currents with the seabed cause significant sediment transport near the shore.

Nonlinear, numerical mild-slope models of random surface wave statistics, including the effects of refraction and diffraction, can explain wave propagation over complex topographies (Janssen et al. 2008). Topography-induced refraction and shoaling in the coastal zone cause the Gaussian distribution of wave statistics to break down by the generation and amplification of wave nonlinearities (Herbers 2003), which in turn have important consequences for coastal circulation and sediment transport (Hoefel 2003). Wave shoaling is induced by a difference of speed between the bottom and the top parts of a wave: while the bottom slows down to satisfy the no-slip condition, the top of the wave moves forward and catches up with the bottom, leading to wave asymmetry, wave skewness, and wave breaking (Elgar & Guza 1985). When waves break, wave energy is dissipated, and what we need to characterise at that stage are the energy dissipation rates and the radiation stresses that induce nearshore currents (Hamm et al. 1993). Except for cases when plunging waves impinge on the sea floor when breaking (Pedrozo-Acuña et al. 2010), generally it is wave-induced currents which are responsible for turbulence generation and for sediment transport at the seafloor bottom, in the surf zone and up to the swash zone.

For waves in the surf zone, the interest is on random (stationary) wave fields, through the velocity potential function, $\Phi(\mathbf{x}, \mathbf{z}, t)$, and the surface elevation, $\eta(\mathbf{x}, t)$, which are Fourier functions of the form

$$\Phi(\mathbf{x}, z, t) = \sum_{p_1=-\infty}^{\infty} \phi_1(\mathbf{x}, z) \exp(-i\omega_1 t) \text{ and} \tag{2.39}$$

$$\eta(\mathbf{x}, z, t) = \sum_{p_1=-\infty}^{\infty} \zeta_1(\mathbf{x}) \exp(-i\omega_1 t), \tag{2.40}$$

respectively. Here $\omega_1 = \omega_{p_1} = p_1 \Delta\omega$, and $\Delta\omega$ is a discrete angular spacing; $\phi_1 = \phi_{\omega_1}$. The wave steepness $\epsilon = a_0 k_0 \ll 1$, and the nondimensional bottom slope $\beta = |\nabla h_0|/(k_0 h_0) \ll 1$, typical of mild-slope seafloors. k_0 and a_0 are the representative wave number and amplitude of the wave field, respectively, and

h_0 and $|\nabla h_0|$ denote a characteristic depth and bottom gradient, respectively. Assuming that $O(\epsilon) \sim O(\beta)$ and that the lowest-order wave–wave interactions are near-resonant, and omitting $O(\epsilon^2)$ terms, leads to a mild-slope equation with a quadratic nonlinear coupling term on the right-hand side:

$$\nabla^2 \varphi_1 + k_1^2 \varphi_1 = i \sum_{\omega_1, \omega_2} W_{23} \varphi_2 \varphi_3 \hat{\delta}_{1;23}^{\omega}, \tag{2.41}$$

where $\hat{\delta}_{1;23}^{\omega} = \hat{\delta}(\omega_2 + \omega_3 - \omega_1)$ is the discrete Dirac delta. The wave number k_1 and the angular frequency ω_1 of a progressive (linear) gravity wave satisfy the dispersion relation $\omega_1^2 = g k_1 \tanh(dk_1)$, and $\varphi_1 = \sqrt{C_1 C_{g,1}} \Phi_1|_{z=0}$, where C_1 and $C_{g,1}$ are the phase speed and the group velocity, respectively, of waves with angular frequency ω_1. If we define $T_i = \tanh(k_i d)$ and $\mathcal{P}_i = \sqrt{C_i C_{g,i}}$ and denote sgn($\cdot$) as the signum of $\cdot$, then the nonlinear term in Equation 2.41 is given by

$$W_{23} = \frac{1}{2\mathcal{P}_{2+3}\mathcal{P}_2\mathcal{P}_3} \left[\omega_2 k_3^2 \left(1 - T_3^2\right) + \omega_3 k_2^2 \left(1 - T_2^2\right) \right.$$
$$\left. + 2 \left(\omega_2 + \omega_3\right) k_2 k_3 \left(\text{sgn}(\omega_2 \omega_3) - T_2 T_3\right) \right]. \tag{2.42}$$

Equation 2.41 is an elliptic equation representing an isotropic wave field, in which the waves indiscriminately propagate in all directions (Janssen et al. 2008). Anisotropic effects, such as wave refraction, backscattering from the seafloor, or reflection from the coast, are not taken into account. In order to include them, some approximation of Equation 2.41 is needed that considers forward scattering. However, the model performs well in the test cases presented by the authors, as the wave height transformation process is well reproduced. They consider a forward scattering approximation of Equation 2.41, with x and y coinciding with the cross-shore and longshore directions, respectively, and the waves propagating in the positive x-plane, with φ_i satisfying

$$\partial_x \varphi_1(\mathbf{x}) = \left(i\tilde{\varkappa}_1 - \frac{\partial \tilde{\varkappa}_1}{\partial x} \right) \varphi_1(\mathbf{x}) + \sum_{\omega_1, \omega_2} \frac{W_{23}}{2\tilde{\varkappa}_1} \varphi_2(\mathbf{x}) \varphi_3(\mathbf{x}) \hat{\delta}_{1;23}^{\omega}, \tag{2.43}$$

where $\tilde{\varkappa}_1 = \text{sgn}(\omega_1)\sqrt{k_1^2 + \partial_x^2}$. For a plane wave over a laterally uniform bottom, $\tilde{\varkappa}_1$ is the principal (x-component) wave number, and the linear part of Equation 2.43, i.e.,

$$\partial_x \varphi_1(\mathbf{x}) = \left(i\tilde{\varkappa}_1 - \frac{\partial \tilde{\varkappa}_1}{\partial x} \right) \varphi_1(\mathbf{x}), \tag{2.44}$$

represents a Wentzel-Kramers-Brillouin (WKB)-type solution, which accounts for the slowly varying depth in the principal direction. The authors do not discuss the coupling of the wave model with sediment transport equations, nor any potential implications of their results to sediment transport and morphodynamics, so we will not discuss this model any further. However, the implications could be assessed by analysing the nearbed velocity field and the bed shear stresses predicted with this modelling approach.

2.3.2.3 Other Current Generation Mechanisms in the Nearshore

Nearshore currents have different origins. Wind-driven currents, for example, are caused by the balance between the wind shear stress at the surface of the ocean and the Coriolis forcing due to the Earth's rotation; this is called *Ekman pumping*. The depth-integrated velocity components of these currents, (U_E, V_E), are of the form

$$U_E = \frac{\tau_{sx}}{\rho f}$$
$$V_E = -\frac{\tau_{sy}}{\rho f} \qquad (2.45)$$

with

$$(\tau_{sx}, \tau_{sy}) = c_D \rho_a |\mathbf{u}_{a10}| (u_{a10}, v_{a10}), \qquad (2.46)$$

where $c_D = 1.03 - 1.35 \times 10^{-3}$ is the sea surface wind-drag coefficient, $\rho_a = 1.22 - 1.25 kg/m^3$ is the wind density, $|\mathbf{u}_{a10}|$ is the magnitude of the wind speed 10 m above MSL, and (u_{a10}, v_{a10}) are the wind speed components 10 m above MSL (Smith 1988).

Ekman processes are also observed closer to the seabed, and they cause currents at the bottom to move upwards across thermohaline layers; this is known as *Ekman lift-up* or a wind set-up (set down) of the water level near the coast.

2.3.3 A Description of Tides

A comprehensive description of tides is well beyond the scope of this book, but tides are an ubiquitous ocean forcing, described as 'the heartbeat of the ocean' by (Defant 1958), a forcing that, although in its fundamental form is fully predictable, is affected by local parameters. Also, the study of tides in combination with storm surges is particularly important in the field of coastal hazard prediction and management. The first step is to discuss the equilibrium theory of tides, followed by the dynamic theory of tides, and then the propagation of tides within estuaries.

Tides are one of the most predictable physical phenomena in the ocean. Tides are long-period, shallow-water waves generated in the ocean by the forces exerted on planet Earth by the Moon and the Sun. At the coastline, tides appear as a regular rise and fall of the sea surface. The period of the rise and fall depends on the dominant tidal frequency at the site. When the crest (trough) of the tidal wave reaches a particular location, a high (low) tide, HT (LT), occurs. The tidal range R_{tide} is defined as the difference in height between high tide and low tide, $R_{\text{tide}} = (\text{HT} - \text{LT})$ heights. A horizontal movement accompanies this rise and fall of the tide, generating a *tidal current*. The world's strongest tidal current is the Saltstraumen Maelstrom, in Norway, while the World's largest tidal range occurs at the Bay of Fundy, in Canada—demonstrating that the sites with highest tidal range do not coincide with the sites with highest tidal currents. In the open ocean, tidal currents are relatively weak, but they become strong near to the coast. There are many local factors that affect the same tidal forcing, including the bathymetry, the orientation, and slope of the coastline, or the dimensions of the water body, for example. In narrow straits and inlets, in very shallow macrotidal regions, or near estuary entrances, tidal currents can reach very high speeds, above 6 km/h (Ross 1995)—as in the 'Mont Saint Michel', in the English Channel, where the tidal range is also very large, of the order of 14 m. Tides may be characterised, principally, by

- the twice-daily or daily variation,

- the difference between the two tidal patterns of the day,

- the spring–neap cycle,

- the annual variation.

The equilibrium theory of tides takes into account not any local effects but only the two main generating tidal forcings: the gravitational force exerted by the Moon, the Sun, and the solar system planets on the Earth and the centrifugal force, i.e., the force on the Earth that is reacting to the gravitational forcing. Taking the case of two interacting bodies for the moment, e.g., the Earth and the Moon, it is well known that the gravitational force, F_g, satisfies Newton's law of gravitation,

$$F_g = G\frac{M_1 M_2}{r^2}, \tag{2.47}$$

where r is the distance between the centres of gravity of the two bodies, M_1 and M_2 their respective masses, and G the gravitational constant.

At the centre of mass of the Earth (CE), the centrifugal force exactly opposes and cancels the gravitational force, so at that point, their sum is exactly zero. However, at other points, this is not so, because the centrifugal force has the same direction and magnitude everywhere, whereas the Moon's gravitational force on the surface of the Earth is always directed towards the Moon's centre of mass, and r is the distance from that point to the Centre of Mass, so the gravitational force is smaller on the points on the Earth's mid-plane that is perpendicular to the axis joining the Earth's and the Moon's centres of mass. The difference between the centrifugal and the gravitational forces is called the tidal *tractive* force. These forces are illustrated in the schematic of Figure 2.15 and explain why the equilibrium theory of tides predicts the generation of two tidal bulges, one on the side closest to the Moon and one on the opposite side. When we 'switch on' the Earth's rotation, we manage to explain the twice-diurnal cycle of the tide that is observed in many places around the globe. The tractive force is very similar in form to the gravitational force but is inversely proportional to the cube of the distance between the centres of mass between the two bodies, and not its square, i.e.,

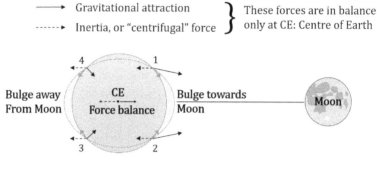

FIGURE 2.15
Schematic of the gravitational force, the centrifugal force, and their difference, the tidal tractive force, at four different points on Earth. (After Fields 2015.)

$$F_{\text{tract}} = G\frac{M_1 M_2}{r^3}. \tag{2.48}$$

The equilibrium theory of tides assumes that an ocean of constant depth all around the globe, that no continents interfere with tidal motion, and that the ocean responds instantly to the tidal forcings. None of this is actually true. Thus, a different, dynamic theory of tides that takes into account these restrictions needs to be developed.

The diurnal tidal inequality is explained by the difference between the Earth's rotating plane and the plane of rotation of the Moon around the Earth. The angle between these two planes is called *the declination angle*. A schematic showing the declination angle, and the diurnal inequality, is shown in Figure 2.16. However, in reality, what is shown in Figure 2.16 does not adhere to observations. Most places have mixed or semidiurnal tides, and there are some places close to the equator, such as the Gulf of Mexico, with diurnal tides. Figure 2.17 shows the characteristics of the tidal pattern at different locations around the globe.

Next, what about the contribution of the Sun? From a house facing East, we see the Moon rising every evening and the Sun rising every morning. Both of them generate fully predictable, tide-generating forces on the surface of the Earth. However, the Sun has 27 million times more mass than the Moon, but it is 390 times further away than the Moon. Let us recall that the tide-generating force between two bodies is proportional to the product of their masses and inversely proportional to the cube of the distance between them. This implies that the tide-generating force on Earth associated with the Sun is about 46% of that associated with the Moon. Also the Moon rotates around the Sun in about 28 days, while the Earth rotates around the Sun in about 365 days.

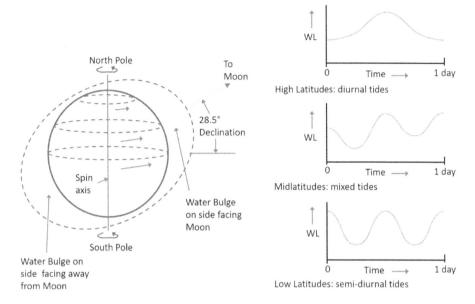

FIGURE 2.16
Schematic of the types of tides predicted with the equilibrium theory of tides. (Based on material from https://slideplayer.com/slide/4742618/.)

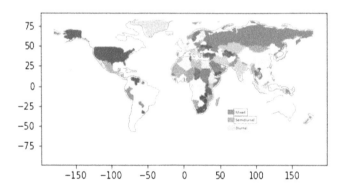

FIGURE 2.17
Observed tidal pattern characteristics around the globe. (Modified from figure in www.seascisurf.com/tides.pdf, courtesy of © 2002 Brooks/Cole.)

So the Sun and the Moon are either aligned with the Earth, in quadrature with the Earth, or something in between. This is shown in schematic form in Figure 2.18 and explains the neap–spring cycle.

At smaller scales, tides can be influenced by a number of local characteristics. For example, the magnitude of tides can be strongly influenced by the

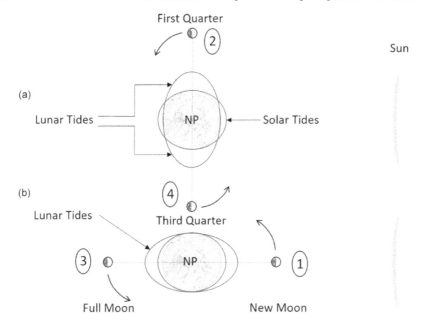

FIGURE 2.18
Joint influence of the Sun and the Moon tides (Earth shown in grey), looking
from the North Pole. (a) During First Quarter and Third Quarter, the Sun
and Moon tides are in quadrature, neap tides observed.(b) During New Moon
and Full Moon, the Sun and Moon tides are aligned, spring tides observed.

shape of the shoreline. Mid-oceanic islands not near continental margins typically experience very small tides, of less than one metre. Conversely, when
oceanic tidal bulges hit wide continental margins, the height of the tides
can be magnified. The shape of bays and estuaries can magnify the intensity of tides. Funnel-shaped bays in particular can dramatically alter tidal
magnitude. The Bay of Fundy in Nova Scotia is the classic example of this
effect (Thurman 1994). Narrow inlets and shallow water also tend to dissipate incoming tides. Inland bays such as Laguna Madre, Texas, and Pamlico
Sound, North Carolina, have areas classified as nontidal even though they
have ocean inlets. Seasonal river flows in the spring can severely alter or mask
the incoming tide (Thurman 1994).

   In order to analyse local effects on tidal propagation, we can perform a
thought experiment, using a pair of small rectangular tanks with different
dimensions, representing ocean basins or funnel-type estuaries. These tanks
are moved back and forth with a period of 1 or 2 s, and the water level and
current speed are then recorded. The tanks will respond differently to the
motion, because each of them has its own resonance period, determined by
the dimensions of the tank and not by the period of the forcing. The one
with dimensions closest to the resonance frequency dimensions will be the one

with strongest water motions. Sloshing motions in moving tanks is an active research area, as demonstrated by the work of Frandsen (2004), Rafiee et al. (2010), or Battaglia et al. (2018), for example.

Tides in the open ocean have very small amplitudes, of the order of 0.5 m at most. However, they can be large in some basins, when the resonant frequency of the basin is close to the frequency of a tidal principal component. The natural period T_n of the basin is the time taken for the wave to leave one boundary and to return after reflection at the second boundary:

$$T_n = \frac{2L}{(gD)^{1/2}}, \tag{2.49}$$

where L is the tank's length and D is the mean water depth. T_n is also called the *fundamental mode of oscillation*, and it has a central location where the water-level change is zero, called a *node*, or an *amphidromic point*. Equation 2.49 is known as Merian's Formula (Pugh 1996).

Standing waves in a basin open at one end can also occur with a double wavelength, and Equation 2.49 in that case becomes

$$T_n = \frac{4L}{(gD)^{1/2}}.$$

An exact quarter wave dimension is very unlikely, but as long as the dimensions of the wave length and a quarter of the basin length are similar, there is a probability of tidal amplification.

Tides in adjacent seas are assumed to be forced, predominantly, by the open ocean. Because of this, tides in adjacent seas propagate as *co-oscillating tides*, because they co-oscillate with the open ocean at their boundary. The tide is then represented as a Kelvin wave, that is, a nondispersive wave that balances the Coriolis forcing of the Earth against a boundary, such as the equator or the coastline. Kelvin waves travelling in opposing directions in a constricted space cause amphidromic points to appear, as discussed above; these points remain at MSL, that is, they suffer from no tidal elevation variation. However, there can be tidal currents since the water levels on either side of the amphidromic point differ. Amphidromic points occur as well in the open ocean. Indeed, whether an amphidromic point can exist in an ocean basin of given dimensions is determined by the Rossby radius of deformation of the tidal wave (Pugh 1996). When the dimensions of the basin are large enough, the Coriolis forcing on the tidal wave causes a full rotation of the wave. As the latitude increases, the Coriolis forcing increases, and the Rossby radius decreases (Reynaud & Dalrymple 2011). The existence of amphidromic points was only speculative even at the end of the 19th century. The fundamentals of tidal theory can be traced back to Laplace, who in 1800 divided tides into diurnal, semidiurnal, and long-period components; this is well described in the history of the theory of tides by Ekman (1993), from the Geological Survey of Sweden. However, it was not until 1868 that William Thomson (later known as

Lord Kelvin) introduced harmonic analysis to decompose tides into harmonic components, and in 1872, he invented the first tide prediction machine (Ekman 1993). In order to be able to identify amphidromes, one would need global cotidal charts of the ocean tides, but the first realistic cotidal map was constructed in 1904 by Rollin Harris (1863–1918), and only when modern sea-level measurements were available, in the 20th century, was it possible to observe amphidromic points in tidal measurements. As highlighted by Pugh (1996), tidal dynamics are not fully understood even today.

2.3.4 Dynamics of Estuaries and Coastal Lagoons

Estuaries may be classified according to their geological origin and sedimentary characteristics, as in Magar (2016), or according to their hydrodynamic and hydrographic properties or a combination of both. Estuaries and coastal lagoons are coastal water bodies where hydrographic and hydrodynamic properties change in relation to the average open ocean conditions, due, for example, to mixing with freshwater coming off the land, temperature and salinity increases due to the shallowness of the water body, or reduced wave energy due to wave dissipation processes at either the mouth of an estuary or the inlet of a the coastal lagoon. The dynamics of estuaries and coastal lagoons vary from submesoscales to macroscales (Hibma et al. 2004). As discussed in Chapter 1, the estuary extent is defined as the landward limit of tidal influence, the location in the fluvial reach, where dynamic processes are affected by tidal action (Perillo 1995). Small- to large-scale bedforms, ranging from ripples to bars, dunes, and sandbanks, are observed in all three reaches: the tidal, the middle, and the fluvial estuary (Seminara & Blondeaux 2001).

Estuaries and coastal lagoons may vary between four different types: well mixed, weakly stratified, strongly stratified, and salt-wedged type (Valle-Levinson 2010). Estuaries change from one type to another based on the balance between the river discharge and the strength of the tides. Strongly stratified estuaries, for example, are a result of moderate to large river discharges and weak to moderate tides; weakly stratified or partially mixed estuaries, on the other hand, are a result of weak to moderate river discharges and moderate to strong tides. Well-mixed estuaries are a result of weak river discharges and strong tides, while salt-wedged estuaries develop during high river discharges and weak tides. Pritchard (1952) showed, for the first time, that the estuarine circulation depended on the horizontal salinity gradient, $\delta S/\delta x$. While tidal currents are typically much stronger, the estuarine (or residual) circulation may be revealed by averaging the vertically varying horizontal currents over a tidal cycle. The salinity gradient, in turn, induces a vertically varying horizontal pressure gradient, $\delta p/\delta x$. This pressure gradient depends as well on the sea surface slope, $\delta \eta/\delta x$,

$$\frac{1}{\rho}\frac{\delta p}{\delta x} = g\frac{\delta \eta}{\delta x} + \beta g\frac{\delta S}{\delta x}(h - z), \qquad (2.50)$$

where ρ is the (salinity-dominated) water density, β is the saline contraction coefficient, g is the acceleration due to gravity, h is the water depth, and z is the vertical coordinate measured upward from the bottom. The force that is most important in balancing the pressure gradient is the internal stress (or momentum flux) acting on the estuarine shear flow. In a tidally averaged sense, generally the estuarine circulation is down estuary near the surface (where the water has less salinity) and up estuary near the bottom (where the water has more salinity), with an inflow of buoyancy and freshwater from the river into the estuary (Uncles & Mitchell 2017).

The turbulent stress, τ [force/area or Pascals], may be expressed in terms of an eddy viscosity, K_z [m^2/s], and the vertical shear of the horizontal flow, $\delta u/\delta z$,

$$\tau = \rho K_z \frac{\delta u}{\delta z}. \tag{2.51}$$

K_z varies between 10^{-4} and 10^{-2} m^2/s in typical estuaries, which is several orders of magnitude larger than the molecular viscosity (Geyer 2010). Cross-channel salinity gradients can only be generated by three mechanisms (of which only two are believed to be dominant): (1) differential advection, which is the development of cross-channel gradients set up by lateral shears in the along-channel flow acting on the along-channel salinity gradient; (2) the tilting of vertical stratification by cross-channel variability in vertical motion associated with the secondary flows themselves; and (3) cross-channel variations in mixing (Chant 2010).

Sediment transport and morphodynamics in estuaries, however, are influenced, mostly, by the relative importance of the hydrodynamical forcings, rather than the hydrographic properties. The along-channel flow presents intratidal and subtidal variability, as well as long-term variability, for example, over the neap–spring tidal cycle (Ross et al. 2019). Intratidal variations are variations at timescales larger than the largest turbulent timescale but smaller than the tidal period (Collignon & Stacey 2013). The intratidal variability is induced predominantly by tidal currents and sea breezes with periods of hours and, in some cases, by harbour or basin seiches with periods of hours to minutes. Vertical mixing, density gradients, tidal straining, and tidal asymmetry are some of the factors affecting estuarine dynamics at intratidal temporal scales. For example, Ross et al. (2019) found that the eddy vertical viscosity (a proxy of vertical mixing) has significant intratidal variability, i.e., between flood and ebb tides. This intratidal variability is of an order of magnitude in fact, and this happens during both neap and spring tides. However, during neap tides, vertical mixing is much smaller than that during spring tides, as expected. Ross et al. (2019) also mention that secondary flows are generated by Coriolis forces and that they play an important role in mid-water mixing processes, in particular during slack waters. The subtidal variability, on the other hand, is affected by the low-frequency components of the tides (with periods of 13–15 days or 27–31 days), low- and high-pressure atmospheric systems (with timescales of 3–10 days), and responses to extreme events or annual flow cycles (with timescales of a few hours or days), amongst others (Jay 2010).

The first analysis of secondary flows is reported in the seminal work by Smith (1976, 1977), on the dispersion of contaminants in small shallow estuaries. The originality of his analysis is based on the use of a frame of reference that moves with the tide, which simplified the interpretation of the dynamics. In small estuaries, the direction of the cross-channel flow may be defined using the channel orientation. The channel orientation can be obtained through a principal component analysis or tidal ellipse analysis of the tidal currents within the channel (Chant 2010), but this requires *in-situ* current meter data along the channel, which may be expensive to obtain. As a consequence of tidal currents being the dominant forcing in estuarine hydrodynamics, the strength of the flow is much smaller (<10%) in the cross-channel than in the long-channel direction. As mentioned earlier in this chapter, when we defined the mobility number (Equation 2.5), the strength of the flow is related to the mobility number ψ, linked in turn to the current velocity. The cross-channel current velocity component is much smaller than the along-channel one. However, the gradients of the flow velocity, and the salinity and turbidity gradients, are much larger than their respective long-channel gradients. Therefore, the cross-channel advective terms in the momentum and the transport equations are larger than their long-channel counterparts. Moreover, it is through the secondary flows that mixing across the channel takes place, and therefore it is critical to include them when estimating transport and dispersion processes in estuaries.

Secondary flows are driven by a number of mechanisms, including the Coriolis acceleration, bed friction, flow curvature, cross-channel baroclinic pressure and density gradients, and diffusive processes in the BBL. Both Coriolis and friction are important, even for estuaries narrower than the Rossby radius. This radius is defined as the ratio of the internal wave speed $(g'h)^{1/2}$ to the local Coriolis frequency f, with $g' = g\delta\rho/\rho$ as reduced gravity, where g is acceleration due to gravity, $\delta\rho$ is the density difference between the upper and lower layers, and ρ is $\approx$1,000 kg m^3. The flow curvature, on the other hand, drives helical lateral flows that are normal to the streamwise flow.

In order to characterise the along-channel and cross-channel dynamics, we follow Chant (2010) and express the governing momentum conservation equations for the streamwise velocity, u_s, the cross-stream velocity, u_n, the pressure, P, and the stress, τ, in curvilinear coordinates (s, n, z), where s is the streamwise direction, n the normal direction, and z the vertical direction:

$$\frac{\partial u_n}{\partial t} + u_s \frac{\partial u_n}{\partial s} - \frac{\partial u_s^2}{\partial R} + f u_s + \frac{1}{\rho}\frac{\partial P}{\partial n} - \frac{\partial \tau}{\partial z} = 0. \qquad (2.52)$$

In Equation 2.52, it is assumed that $u_n \ll u_s$, which, as we know, is a valid assumption, because of the predominantly long-channel propagation of the tidal currents.

There is a positive feedback between secondary flows and channel morphology, as highlighted in recent work by Huijts et al. (2006) and Fugate et al. (2007). As mentioned by Chant (2010), depth-averaged cross-channel flows will develop in the presence of long-channel morphological gradients, but if there are no long-channel morphology changes, then one observes closed circulation cells with zero cross-section-averaged flows. In contrast, with long-channel morphological changes, a cross-sectional flow develops, driving a lateral accumulation of sediments. These lateral flows are influenced by Coriolis forces and density gradients, and the rate of erosion, transport, and deposition of sediments depends on the bed shear stresses and sediment availability (Huijts et al. 2006).

2.4 Seafloor Composition

As in the case of the dynamical characteristics of coastal environments, seafloor composition in continental shelves generally is different from sediment composition in estuaries, deltas, or tidal coastal lagoons and will hence be considered in two separate sections. Although sediment sizes cover a large spectrum, they can be divided broadly into three basic types: cohesive, pseudocohesive, and noncohesive sediments, based on their median grain diameter, d_{50}. Cohesive and pseudocohesive sediments have a d_{50} of 63 μm or less, while noncohesive sediments have a d_{50} above 63 μm. Cohesive sediments include clays and muds with a d_{50} of 2 μm or less, while pseudocohesive sediments include silts and muds, with 2 μm $< d_{50} <$ 63 μm. Noncohesive sediments, on the other hand, may be divided into sand, with 63 μm $< d_{50} <$ 2,000 μm, and gravel, with 2,000 μm $< d_{50}$. On open shores subject to wave action, generally sediments are noncohesive except, perhaps, in areas with some vegetation coverage in tropical regions, such as mangroves on exposed coastlines. Cohesive sediments are found in estuaries, coastal lagoons, and river deltas.

2.4.1 Sediment Composition on Open Shores

Typical sediments on a continental shelf area are depicted in the diagram of seabed sediment types published by Mengual et al. (2017). Although sediments cover the full size spectrum, i.e., from large rocks to clays, the percentage coverage of the seabed for the different particle sizes is very different. Figure 2.19 shows, with a resolution of about 2 km, the seabed composition off Le Croisic, on the northeastern side of the Bay of Biscay and on the Western Approach to 'La Manche', the English channel, at depths from around 140 m to the coast. The first characteristic feature of the region is that between the 100 and 130 m isobaths; the seabed is composed mostly of sand and silt, with the exception of a small area below 46° 30′ N where the seabed is composed mostly of gravel and stones, but it is right at the edge of the domain so it is difficult to assess the extent of this patch. The second aspect is that, in the region shallower than the 130 m isobath, a variety of sediment compositions occur, that may be analysed in terms of latitude and depth limited regions: south of 46° 40′ N, a patch of sand mixtures is found; north of 46° 40′ N and south of Le Croisic are sand mixtures between 40 and 100 m depths, gravel mixtures between 40 m and outside the Le Loire Bay, and sands and muds in the Loire estuary and Loire Bay. North of Le Croisic and at depth below the 100 m isobath, the floor composition is very variable, with patches of muds in the Vilaine Bay and the North West area of the domain, while the rest of the seabed is gravelly with very small patches of sand mixtures.

Figure 2.19 is quite unique in the level of sediment composition detail, which allows a modeller to build roughness coefficient maps that can be fed into numerical models to assess the impact of seabed roughness on the regional dynamics. Such models can then be used to determine maps of bedload mobilisation functions. The seabed composition map of Great Yarmouth, shown in Figure 2.20, has similar level of resolution to Figure 2.19 but is less detailed

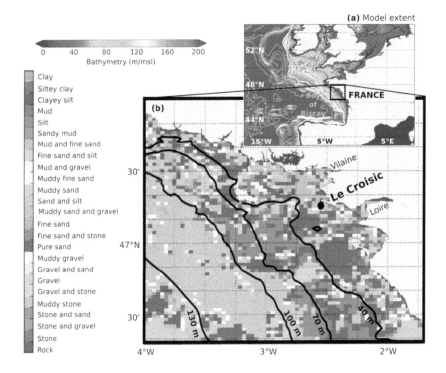

FIGURE 2.19
Coastal seafloor composition off Le Croisic, typical of 'La Manche', the English Channel. (From Mengual et al. (2017)—reproduced with permission.)

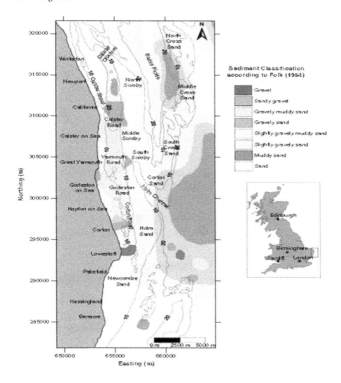

FIGURE 2.20
Coastal seafloor composition off Great Yarmouth, typical of the North Sea.
(From Reeve et al. (2008)—reproduced with permission.)

because only eight sediment classes have been used, in contrast with 25 classes used in Figure 2.19. Using so many different sediment composition classes may not be so relevant for regional morphodynamics modelling but could be important for identifying seabed locations that are best for offshore engineering activities.

The seabed roughness k_s, depends not only on sediment composition, but on also on any bed features, such as ripples or dunes (Davies & Robins 2017). Seabed roughness alters the friction velocity, hence bed shear stress, and hence the sediment transport rates near the seabed. Usually the bedforms have scales which fall within subgrid scales, and therefore their effect on k_s needs to be parameterised. Davies & Robins (2017) have shown that a model with spatially and temporally varying roughness can provide a realistic description of the bedforms and also reproduce the tidal flow with good accuracy. Van Rijn (2007) formulated a bottom roughness based on noncohesive sediment and bedform-dominated roughness that has been implemented in some regional models, such as TELEMAC.

In any case, it actually is not always possible to have access to seabed maps with good accuracy at a resolution of the order of one kilometre over a relatively large domain, so necessarily the number of sediment classes into which the sediment composition is divided is also small. For example, in Figure 2.21 for the Upper Gulf of California, although the spatial resolution is again of the order of a few kilometres, with the spatial extent of the domain being around 380 by 280 km, there are now only six sediment classes, which is 20% less than the number of classes in Figure 2.20. This resolution is very coarse for certain applications, for example, for local models requiring a spatial resolution of the bathymetry and the seabed composition in the order of tens of meters, such as coastal flooding issues at the scale of a town or city, where we are interested in assessing which neighbourhoods will be most affected by a coastal storm. For mesoscale modelling, however, this resolution may be fine provided that one is only interested in monthly to yearly behavioural trends. So for local models, ideally one would have access to bathymetric data from multibeam surveys, or to satellite-derived bathymetries with good resolution. Then one has the option to use a coarse grid and smooth the bathymetry when interpolating the data onto that grid or use a fine grid that reproduces the small bed features captured by the high-resolution sampling of the survey.

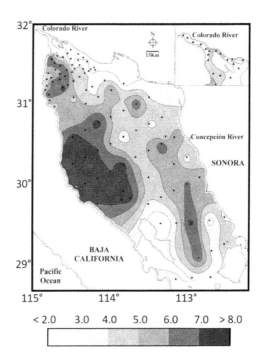

FIGURE 2.21

Coastal seafloor composition in the Upper Gulf of California. The sediment grain size is shown in Φ units, where $\Phi = -\log_2(d_{50})$; d_{50} is the median grain diameter in millimetres. (From Carriquiry et al. (2001)—reproduced with permission.)

2.4.2 Sediment Composition in Estuaries

Estuarine sediment composition varies with tidal range, river discharge, wave height (near the mouth), sediment source (either marine or alluvial), and sediment availability, which all affect sediment transport and morphodynamic processes. All these estuarine forcings have an axial, lateral, and vertical impact (Nichols & Biggs 1985). Towards the head of the estuary, bed composition includes silts, clays, plants, and roots, grading down to sand, gravel, and cobbles. Large clay and silt deposits may be found towards the mixed region of the estuary, together with sandy lenses and laminae. Towards the estuary mouth, the main sediments are marine sands with abundant cross-bedding, and tidally driven sandbanks may be found here. These sandbanks have a low-angle cross-bedding in fine sands with silt laminae. Lateral variations also occur, with shorelines composed of sand, gravel, shells, muds, plant fragments, and basal peat; subtidal flats composed of laminated muddy sands and sandy muds; and mid-channel environments dominated by coarse marine sands and massive cross-bedding.

It is possible to analyse sediment composition according to an estuary's classification, which includes coastal plain estuaries, rias, fjords, delta front estuaries, structural estuaries, coastal lagoons, and others. A more complete category list may be found in Perillo (1995), who also divided estuaries into primary and secondary estuaries. Primary estuaries have not changed much since their genesis, while secondary estuaries have been modified significantly from their original form by changes in sediment availability and by hydrodynamic forcings. Three reaches may be identified in estuarine physiography with different forcing and sediment characteristics. Cohesive silts and clays in the upper reaches (near the estuary's head) have an alluvial origin, while noncohesive sands, gravels, and cobbles in the lower reaches (near the estuary's mouth) have a marine origin. Large clay and silt deposits may be found in the middle reaches of the estuary, together with sandy lenses and laminae. Sediment composition in an estuary also varies laterally, comprising, for example (Nichols & Biggs 1985), shorelines of mixed muds, sands, and gravels and organic components such as shells and plant fragments; subtidal flats of muddy sands or sandy muds; and mid-channel environments of coarse marine sediments.

Figure 2.22 shows a map of the interpolated median grain size (in meters) in the Bay of Fundy, with some geographical annotations. From this map, it can be seen that, in the Upper Bay of Fundy located East of longitude 65.5° W, the sediment is composed of sands with median grain diameters mainly between 0.5 and 2 mm or noncohesive fines. The Lower Bay of Fundy is approximately defined as the region between the mouth and longitude 65.5° W. Here the 'mouth' is assumed to be a transect approximately parallel to the Northern end of Grand Manan Island, cutting across Grand Manan basin, touching the southern tip of Brier Island, and reaching towards Yarmouth (43° 50′ 14.66″ N, 66° 6′ 43.93″ W)—see Figure 2.22 (Li et al. 2015). The Lower Bay of Fundy has two distinct sedimentary regions: a 'muddy region' with grain

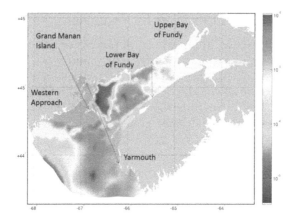

FIGURE 2.22
Sediment composition in the Bay of Fundy (mean grain size [metre]), with the three sedimentary region boundaries discussed in the text. (From Li et al. 2015—reproduced with permission.)

sizes <100 μm and a 'gravelly region' with fines and gravels up to centimetre scales. The muddy region is on the Bay-facing side of Grand Manan Island and covers an area of roughly 1,500 km². The gravelly region covers the rest of the Lower Bay of Fundy. The third sedimentary region is the Bay of Fundy's western approach. The region east of 67° W is composed of muds towards the sea, and fines and gravels in the rest of the region, with a homogeneous area with grain sizes of $1 - 3 \times 10^{-3}$ m near the coast. The region west of 67° W is mostly gravelly. The seabed sediment composition in estuaries covers a large range of sediment sizes, from fines to coarse materials, and sediment composition also varies spatially depending on the hydrodynamic conditions and the location along the estuary.

An important characteristic of estuaries and some coastal lagoons is that some regions present a large mud content, originating from the rivers that flow into the estuaries. The areas with large muds are areas where the flow speeds are lower so the fines settle onto the bed or areas with some materials, such as organic dissolved matter, which increase the cohesiveness of the bed. This cohesion keeps fines on the seafloor, even during high river discharge events or high tidal flows. Muds are less common in open seas, except when such muds are ejected into the ocean through the seafloor, generally at high temperatures.

2.4.3 Tidal Flats

Estuaries and coastal embayments in mesotidal to macrotidal environments are characterised by tidal flats with a significant socio-ecological and economic value, in particular because they provide protection against flooding and erosion (Van Maren & Winterwerp 2013). Tidal flats connect the land and the sea and, therefore, are found at the head or the mouth of estuaries; between channels and salt marshes in temperate regions; and between channels and mangroves in tropical regions. Tidal flats are composed of fine-grained sediments, which may have cohesive or noncohesive compositions. Tidal flats composed of noncohesive sediments have been extensively studied and are generally well understood, as opposed to their cohesive counterparts.

The simplest model of intertidal flat morphology was proposed by Friedrichs & Aubrey (2013), using the concept of hypsometry. Hypsometry was formally introduced to the field of geomorphology by Strahler (1952). It quantifies the distribution of elevation or relief from a base level across a drainage basin, through an hypsometric curve (HC), and the cumulative horizontal basin area as a function of elevation, through an hypsometric integral (HI). This simplest intertidal flat morphology model is based on the assumption that the maximum tidal velocity magnitude,

$$U = \frac{\pi L}{T}, \ x \leq \frac{L}{2},$$ (2.53)

is proportional to the ratio of the horizontal distance from the low to the high water line, L, and the tidal period, T (see Schematic in Figure 2.23). Friedrichs & Aubrey (2013) showed that convex hypsometry is correlated with large tidal ranges, low wave activity, or long-term accretion. In contrast, concave hypsometry is correlated with either low tidal ranges, high wave energy, or long-term erosion. The relationship in Equation 2.53 assumes that tidal flats are in equilibrium when the maximum bottom shear stress is spatially uniform. This is not true for tidal flats which have linear slopes away from the shoreline, as such bed profile would not have spatially uniform bed shear stresses, neither under tidally dominated environments nor in wave-dominated ones. When tides are dominant, the action of tidal currents on the seafloor produces a convex hypsometry, and when wind waves are dominant, their

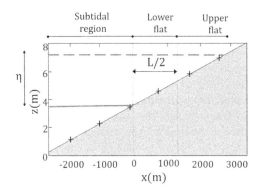

FIGURE 2.23
Idealised tidal flat. (Modified from Hsu et al. 2013.)

action on the seafloor produces a concave hypsometry. Shoreline curvature also has a significant effect on coastal hypsometry, which can be as important as the effect of maximum bottom shear stress. In embayed beaches, for instance, a convex equilibrium hypsometry is more likely to be observed, while a lobate beaches favour a concave hypsometry. This is to be expected, as at embayed beaches, wave action is generally low, while lobate beaches are a point of wave focusing (through refraction effects) and hence a point of enhanced wave energy. At the upper tidal flat, U has a different expression of the form

$$U = \frac{2\pi}{T}L\left(\frac{x}{L} - \frac{x^2}{L^2}\right)^{1/2}, \ x > \frac{L}{2}.$$

These expressions are very useful, as they also enable an estimate of the bottom shear stresses on a tidal flat, together with an understanding of the morphology of many tidal flat systems (Hsu et al. 2013).

Intertidal mudflats generally behave as follows. During calm conditions, tidal forcings alone can drive landward transport and accretion of the upper flat. However, during storm conditions, storm waves may cause significant resuspension of sediment, driving seaward sediment transport, and subsequent erosion of the mudflats. Previous researchers have found that tidal velocity skewness, together with phase differences between tidal velocity and tidal level fluctuations (the Stokes drift), causes a tidally averaged residual transport of sediments (Uncles & Jordan 1980, Feng et al. 1986). Tidal asymmetry also influences the dominant direction of net sediment transport, with dominant flood (ebb) tide causing net landward (seaward) transport (Uncles 1981, Christie et al. 1999). The processes generating residual transport mentioned so far are very local processes, and therefore in some regions, sediment transport will occur landwards, and in other regions of the same estuary, it will occur seawards. A spatially and temporally varying bed roughness also has effects on the residual transport of sediment, as investigated by Davies & Robins (2017). At the Menai Strait, a tidal channel between mainland Wales and the Island of Anglesey which exhibits strong tidal flows and a residual-induced flow, they concluded that a model with spatially varying bed roughness accurately predicts the average residual flow over the neap–spring tidal cycle. The average residual flow can be recovered with a constant, 'effective' bed roughness that is less than half the maximum bed roughness observed on the mid-channel dunes but around 50% larger than the spatially averaged bed roughness. The effective bed roughness is approximately equal to the height of the dunes along a mid-channel transect, but one would have expected it to be half the height. The difference is related to the change in shape of the dunes during flow reversal, when the dune wavelength is shorter than that of dunes in equilibrium.

A nonlocal process also causing residual sediment transport is the settling-lag effect, which gives a net transport in the direction of lower maximum bottom stress (or tidal velocity). As explained previously, suspended sediments settle at a finite speed out of suspension, and thus they are carried landwards on the flood tide (or wave run-up) for some distance after the local fluid velocity has fallen below the threshold velocity for deposition. Then these sediments are re-entrained on the ebb tide (or wave run-down)by a fluid element located further landwards than the fluid element that deposited them. When utilising the classic theory of Friedrichs & Aubrey (2013), the settling lag effect can explain accretion in the upper flat (Chen et al. 2010), for example. In order to quantify properly the contribution to sediment transport associated to the settling-lag effect, it is necessary to carefully consider all sediment transport processes causing suspension, advection, and deposition. The resuspension and advection depend strongly on the dynamics at the turbid tidal edge, the very shallow region near the land–water interface at both ebb and flood tide. It is important for numerical models to resolve all these sediment transport mechanisms, in order to identity the contribution of each of them to the overall transport of sediments in tidal flats. The sediment transport associated with these effects needs to be parameterised in regional models, because the computational cells are much larger than the sediment particles. Models that have robust numerical schemes for the wetting and drying effects would be more likely to capture the settling-lag effects, providing that they consider the dynamics of the coastal bore forming once waves break and that this is coupled with the sediment transport and morphodynamic modules.

2.4.4 Continental Shelf Bedforms and Coastal Features: A Description

Continental shelf bedforms may be of only a few centimetres to tens of metres high. Bedforms may be classified as ripples, megaripples, sandwaves, dunes, shoals, or sandbanks, depending on their size. Shoals are also known as linear sandbanks or tidal sand ridges (Sanay et al. 2007). In some other cases, there are overlaps between definitions, so although we will attempt to provide some sort of definition for different bedform types, we will also try to discuss generic features that apply to all bedforms, at different temporal and spatial scales, as suggested by Winter (2006).

The formation of ripples has been discussed to some extent in the previous section, where the microscale processes of sediment transport at the bed level were discussed. While subaqueous ripples are only a few centimetres high, subaqueous river megadunes are in the order of 10 m high and in the order 100 m long. Subaqueous barchan dunes are sort of lunate bedforms, which may be observed either in isolation or in clusters. In contrast, transverse dunes cover large areas and resemble ploughed fields (Charru et al. 2013). Ripples and dunes also form in desert landscapes through wind action on the sandy bed, but aeolian dune may be between 10 m and 1 km long, so generally, they are an order of magnitude larger than their subaqueous bedform equivalents. This is an interesting observation, but it is even more interesting to understand the bed evolution factors that depend on the environment and the bed materials. It is also interesting to use methods that allow us to generalize some observations, for example, through static and dynamic dimensionless parameter considerations.

Bedform formation and evolution, generally referred to from now on as *bedform dynamics*, may be influenced by tidal currents, marine currents, or storms. Coastal bedforms are also influenced by the action of winds. Some of the earliest research on coastal dunes goes back to the 18th century. Claude Masse, who drew the maps of the Médoc littoral, provided in his 1732 memoirs an instructive description of the Médoc dunes and their evolution (as cited in Buffault 1942). However, the first known manuscript that describes the Médoc dunes is a text written in 1580 by Montaigne (as cited in Buffault 1942), which mentions the approximate area covered by the dunes, their marine origin, and the effects of the ocean and the winds on their inland progression. To my knowledge, this is one of the earliest coastal dune accounts. Research on bedform dynamics has been ongoing, with open questions still to be answered even today. Several researchers since the second half of the 20th century to date have focused their efforts on this topic. The mean currents driving seabed evolution, however, have been reported in studies from as early as the 1847 study by Stokes (1880) (and republished in 1880), but he only considers the inviscid case, and as M. S. Longuet-Higgins (1953) highlights, the viscous terms need to be taken into account in order for the nonlinear components of the wave and their propagation properties to agree with experiments. M. S. Longuet-Higgins (1953) also showed that the steady increase of the velocity from the bottom to the surface and the zero velocity gradient at the bottom are both consequences of the irrotationality of the flow and not of the smallness of the wave amplitude as assumed by Stokes (1880).

In flows with negligible rotation, the Coriolis parameter, f, is much smaller than the wave frequency, σ: $f \ll \sigma$. When rotation is negligible, the oscillatory current in the BBL determines the characteristics of the current just above the BBL, but when $\sigma/f = O(1)$, that is, for flows in rotation, the Lagrangian mean velocities are geostrophic (Moore 1970). The asymmetries in the flow generate residual currents, as is the case for tidally driven sandbank evolution. The effect of such residual currents on the evolution of marine sandbanks is discussed by Huthnance (1973), using the Norfolk marine sandbanks as case study.

Huthnance (1973) used a model for a homogeneous fluid, with horizontal velocity components $(u, v) = \vec{u}$ in the horizontal (x, y) directions, between a fixed bottom $z = -H(y)$ and the free surface $z = \zeta$, $z = 0$ being the MSL. The system rotates with a uniform Coriolis parameter f. The diagram in Figure 2.24 depicts the different parameters and variables in the model. The model assumes a balance between three forces: inertia (combining temporal gradients and convection), rotation, and friction. If (u, v) are assumed to be uniform in x, the depth-averaged continuity equation is of the form

$$\frac{\partial \zeta}{\partial t} + \frac{\partial}{\partial y}\left[(H + \zeta)\, v\right] = 0, \tag{2.54}$$

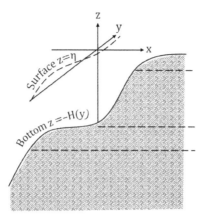

FIGURE 2.24
Shallow-water model with a drag force of the form $-\rho k \vec{u}$. (After Huthnance 1973.)

while the (also depth-averaged) momentum conservation equations can be written as

$$\frac{\partial u}{\partial t} + v\frac{\partial u}{\partial y} - fv = -\frac{1}{\rho}\frac{\partial p}{\partial x} - \frac{ku}{H} \tag{2.55}$$

$$\frac{\partial v}{\partial t} + v\frac{\partial v}{\partial y} + fu = -\frac{1}{\rho}\frac{\partial p}{\partial y} - \frac{kv}{H} \tag{2.56}$$

after averaging in depth and assuming uniformity of (u, v) in x. Analytical solutions of equations 2.55 and 2.56 may be computed under a number of assumptions, as discussed by Huthnance (1973). The analysis is based on the following:

- Topography and currents are independent of x;
- frictional effects are represented by linear bottom drag $-\rho k\vec{u}$;
- continuity and momentum equations are depth averaged and, therefore, in nonconservative form (this means that the form of the acceleration terms has been altered);
- the convective term in the continuity equation is much larger than the (temporal) gradient of the sea surface elevation;
- $\left(\dfrac{H}{L}\right)^2 \left(\dfrac{v}{|\vec{u}|}\right)^2 \ll C_D^2$, or $\dfrac{\sigma H C_D}{|\vec{u}|}$;
- u_∞ and v_∞, the free stream velocity components, which provide the boundary conditions, are sinusoidal and in phase (or antiphase).

U and L are typical velocities and lengthscales, respectively, and C_D is the bottom drag coefficient.

In this model, f is uniform and the Lagrangian mean velocities will follow the depth contours (Huthnance 1973), and since the Lagrangian velocities are responsible for sediment transport, sediment transport will also follow the depth contours. This intense research on sandbank dynamics, led Huthnance (1973) to conclude the following. He first identified that tidal asymmetry is very important for sandbank formation, because it induces a net current or circulation in their proximity, which in turn induces a net bedload transport. Moreover, Coriolis effects force the Lagrangian mass transport velocity to be geostrophic; that is, the pressure gradients are in balance with the forces induced by the rotation of the Earth. Rotation may be considered negligible when the Coriolis parameter, f, is much smaller than the frequency of the wave, σ. When $f \ll \sigma$, the BBL determines the characteristics of the mean current above, as shown by Huthnance (1973). For the case of the Norfolk sandbanks, these forces generated the clockwise circulation of a mean current around the sandbanks.

The early studies on sandbank dynamics by Huthnance (1973) focused on the linear and nonlinear processes driving sandbank evolution. Current research on this topic is focused, in contrast, on bedform–shoreline systems, for example, on the effects of nearshore sandbanks on shoreline evolution (Hequette & Aernouts 2010); the interactions of shorelines with multiple nearshore sandbar systems in natural and nourished beaches (António 2017); the interactions between subtidal mobile sandbanks and intertidal sands mudflats (Elliott et al. 1998); or tidal flats and salt marsh interactions (Mariotti & Fagherazzi 2010), to name but a few.

Of particular importance are those nearshore and coastal features considered to be special areas for conservation (SACs), for example (Elliott et al. 1998),

- subtidal mobile sandbanks
- intertidal sand and mudflats
- estuaries
- large shallow inlets and bays
- lagoons
- reefs,

which, moreover, may dynamically and hydrographically influence the coastal environment around them. All of these seabed and coastal physiographic features have sections that are affected by the incoming waves and other sections where other dynamic forcings are prevalent, such as tidal currents, density gradients, sediment pick-up and deposition, vegetation cover and type of vegetation, river discharge, and the nature of the bottom substratum. Models coupling some of these forcings are more complex but lead to a better understanding of coastal *ecogeosystems*, systems where ecological and geophysical mechanisms interact with each other. Intertidal mudflats are located in low-energy, sheltered areas, whereas intertidal sand flats have an exposure gradient (from low- to high-energy exposure). Intertidal flats are covered for part of the tidal cycle. Subtidal flats and mobile sandbanks are covered throughout the tidal cycle. Subtidal mobile sandbanks tend to have no or very little vegetation. However, salt marshes, kelps, or mangrove forests, for example, may cover the sands and mudflats in estuaries, coastal lagoons, or sheltered bays. These ecosystems coevolve with the tidal flats and the vegetated platforms (Mariotti & Fagherazzi 2010), with vegetation regulating erosion and sediment trapping and erosion (Hir et al. 2007), as well as wave dissipation (Möller 2006). And once a scarp between the vegetated platform and the tidal flat has formed, then local erosional processes modify the rate of regression of the vegetation, ultimately causing its disappearance all along the coastline (Mariotti & Fagherazzi 2010).

As mentioned above, subtidal mobile sandbanks have little vegetation, and are composed of coarser material than sand and mudflat, with high median diameter, low sorting coefficient, high permeability, generally high porosity (depending on compaction), and low sediment stability. Depending on how energetic the environment is, mobile sandbanks may be stable or unstable. Subtidal mobile sandbanks develop in medium- to high-energy environments, as a result of an interaction between the physiography and submesoscale to mesoscale current patterns producing gyres. Some sandbanks, in fact, may be sinks of material that falls to the seabed at the centre of the gyre, including those that are of anthropogenic origin, such as plastic waste.

Extreme events have important effects on both intertidal and subtidal bedforms and on vegetated seabed platforms. Extreme events cause high-energy conditions, causing sediment mobilization and beach erosion, as well as temperature and salinity disturbances in shallow areas, in estuaries, and in coastal lagoons. Climate change impacts such as sea-level rise and tidal elevation changes are going to produce coastal squeeze when the upper tidal boundary is restricted, either by natural or man-made coastal structures, increasing the rate of erosion of the coastline. Organisms may also be affected by global warming and changes in high-temperature extremes, altering their diurnal and their reproductive cycles, disrupting them to possibly dangerous levels. Increases in storminess will affect freshwater run-off and salinity gradients in estuaries. Any stressor causing changes in the shore topography will alter biophysical processes. Storms with one in 25 or one in 50 years return period may cause significant erosion, by turning over the sediments and causing damage to the benthic organisms on the surface of the sandbanks. According to Dyer (1997), sandbanks in estuaries and in nearshore regions are subject not only to extreme events but as well to erosion–deposition cycles, reflecting periods of highest–lowest hydrodynamic energy, such as spring–neap cycles, or winter–summer cycles. Mobile sandbanks need to be subject to strong currents in order to increase the shear stresses on the seafloor. Clearly, most of the sediment is moving the surface of the sandbanks and producing the visual effect of complete sandbank translation. Such sediment motion causes generation and migration of megaripples on the surface of the sandbank (García-Hermosa et al. 2009).

Shallow inlets and bays, by definition, are bounded by a hard substratum and have a soft substratum in the body itself. Their shape is easy to identify, but their size not so, because the seaward boundary is difficult to define. Most of them are saline environments with little river run-off. The forcings that dictate the sedimentary patterns of behaviour depend on whether these are low- or high-energy environments. In high-energy environments, the wave climate, the direction of the prevailing currents, and the length of fetch will affect the structure of the seabed and the type of habitat, while in low-energy conditions, the tidal, the wind-induced, and the residual currents are all of importance. The depth of the feature is determined by the geology and the bathymetry, as well as the tidal regime, defining intertidal zone (the part of a shore between the high water and the low water), infralittoral zone (the region of shallow water closest to the shore, excluding the intertidal zone), and circalittoral zone (the region of the sublittoral zone which extends from the lower limit of the infralittoral to the maximum depth at which photosynthesis is still possible). Tidal inlets and shallow bays are also affected by extreme events, such as storm waves and storm surges (Irish & Cañizares 2009), but storms may have a stabilizing effect on tidal bays (Castagno et al. 2018), contrary to what one may have assumed, by providing the sediment necessary to counteract sea-level rise. However, it is important to bear in mind that inlets do not seem to affect the storm hydrographic measurements used to drive morphodynamic models, at least for inlets with an idealised profile and a width of up to 1,000 m (Salisbury & Hagen 2007). This has important implications for large-scale model set-ups, which in some cases simplify the coastline by ignoring the inlet-bay systems: such models are used for generating boundary conditions for local inlet-bay dynamics or bridge scour models, allowing for the spatial variance of storm surge hydrographs to be captured by the local circulation model.

2.5   Morphodynamics

Morphodynamic processes in coastal environments, as discussed in previous sections, are driven by both small-scale and large-scale mechanisms. At small scales, the dominant processes involve turbulence and vortical motions, and at large scales, the forcings may include plunging breakers, low-frequency waves, coherent vortices, and wave- and wind-induced currents (Hamm et al. 1993). In this chapter, we have described the hydrodynamic and sediment transport processes of most relevance, but some of these need to be simplified such that numerical models are both computationally efficient, while their predictions remain sufficiently accurate. For example, Hamm et al. (1993) mention a number of pointers that need to be considered:

- Wave directionality and randomness need to be included in morphodynamics modelling;

- energy dissipation needs to be modelled accurately all the way up to the swash zone, because wave dissipation induces radiation stresses and low-frequency waves driving sediment transport;

- at tidal inlets, river mouths, and coastal structures, refraction of waves over locally and horizontally sheared currents is important;

- wave nonlinearities grow as waves propagate and break over bars and shoals; hence some parameterizations of nonlinear processes need to be included;

- nonlinear effects are crucial to capture wave shoaling processes—as mentioned by Elgar & Guza (1985);

- infragravity (low-frequency) waves play a crucial role in sediment transport and in the shaping of beaches.

Waves affect only the seaward face of SACs, while in more sheltered areas, tidal currents are more important (see previous section). On longer timescales, other factors may influence morphodynamic evolution; for example, on interannual scales, global-scale phenomena, such as the North Atlantic Oscillation (NAO) in the Atlantic or the El Niño Southern Oscillation (ENSO), could also be driving seaward–shoreward migration of some coastal bed features, such as sandbars (Ruggiero et al. 2005, Magar et al. 2012). However, the annual signal is also prevalent in the wave climate (Magar et al. 2012), as in the NAO, making it difficult to determine whether the bed is responding to NAO or to the waves or to both. This highlights that when two physical phenomena share a temporal cycle, there is not necessarily a cause–effect link between them. When studying a remotely generated forcing, it may be difficult to understand its behaviour unless we use global models that can capture that forcing properly. Not only is it a challenge to model correctly the wave and the circulation conditions for some specific cases, but also the coupling between the hydrodynamic conditions and the morphology evolution equation can be implemented in different ways, which can have an impact on the model predictions.

In advection–diffusion problems, some models are implemented in such a way that the hydrodynamics are computed first and then the advection–diffusion equation is solved, either in an Eulerian or in a Lagrangian framework. For some applications, the bed may be assumed to have reached an equilibrium state, and only the sediment dynamics are analysed, as in (Malarkey et al. 2015). However, in many situations, and hence in many morphodynamic models, the seabed evolves with time, and such evolution needs to be included in the system of conservation equations. The seabed evolution is computed using Exner equation (Exner 1920, 1925), originally developed in river morphology studies, expressed in its original form as (Paola & Voller 2005)

$$\frac{\partial z}{\partial t} = -A\frac{\partial U}{\partial x},$$
(2.57)

where z is bed elevation relative to some fixed datum, t is time, A is a coefficient, U is average flow velocity, and x is downstream distance. The bed elevation fluctuations have different scales: at the finest scale, the fluctuations correspond to the d_{90} of the bed material (that is, the largest grain diameter of ninety percent of the material), to the coarsest scale of sand mounds and river bends, sandbars, dunes, ripples, and their interactions. Equation 2.57 expresses a balance between the local bed elevation changes, on one side, and fluctuations in the rate of sediment transport, on the other (Parker 2008, Davies & Thorne 2008). Exner equation is of the form

$$\frac{\partial z_b}{\partial t} = \frac{1}{1-n}\nabla \cdot \mathbf{Q}_T$$
(2.58)

when expressed in terms of the bed porosity, n, the seabed elevation z_b, and gradients of the total sediment transport rate vector, $\mathbf{Q}_T$—including bedload and suspended load. The suspended load, however, needs to be determined carefully, as it interacts minimally with the seabed and therefore contributes less than the bedload to the bed-level changes. The deposition (D) and erosion (E) sediment flux terms in the sediment concentration (C) balance equation, for example, should be computed not at the seabed, but at a height $z = a$ corresponding to the interface between the bedload layer and the suspended load layer (Davies & Thorne 2008):

$$(E - D)_{z=a} = w_s \left(C_{eq} - C_{z=a} \right),$$
(2.59)

where w_s is the sediment settling velocity, C_{eq} is the equilibrium concentration expected from local bed shear stress considerations at $z = a$, and $C_{z=a}$ is the actual concentration inferred from the upstream sediment flux.

A generalized mass balance formulation of Exner equation was postulated by Paola & Voller (2005), which allows for the evolution of the interfaces between bedrock and sediment and between sediment and water flow, in an independent manner. In this section, we will try to synthesise the analysis of (Paola & Voller 2005). The details may be consulted in the original paper. Their study is based on a three-layer system consisting of a bedrock layer (layer 3), a sediment layer (layer 2), and a flow layer (layer 1). The bedrock is assumed to have constant mass and density, but this assumption can be relaxed to include changes in density and mass gain or loss, as for the other layers. Layer 2 is the sediment layer. This layer is mobile, but it evolves at a much slower rate than layer 1. In fact, only layer 1 is considered to evolve rapidly. These assumptions are in agreement with those in most coastal morphodynamic models. Layer 1 is the water layer, where there is some sediment in suspension, with volumetric sediment concentrations, $c_s = \alpha_s/\rho_s$, which are much smaller than 1, $c_s \ll 1$. Here ρ_s is the sediment particle density, and α_s is the bulk density of the deposited sediment. Note that, in the sediment layer, $c_s = O(1)$.

The three-layer system sketched in Figure 2.25 provides definitions of the following parameters and variables: the elevation η is the conventional topographic elevation, referred to the surface of the Earth; η_b is the bedrock elevation; Ω_{0b} is the mass flux across interface $0b$ (positive for the top layer to the bottom layer, by convention); $\Phi_I = \int_{\eta_i}^{\eta_{i+1}} \alpha_I \vec{v} dz$ is a line flux in the sediment layer $I = s$ or in the flow layer $I = f$; and h_f is flow layer thickness.

The mass conservation equation in layer I can be expressed as

$$\frac{\partial}{\partial t} \int_{\eta_i}^{\eta_{i+1}} \alpha_I dz + \vec{\nabla}_H \cdot \vec{\Phi}_I - \Omega_{\text{in}}(\eta_{i+1}) + \Omega_{\text{out}}(\eta_i) - \int_{\eta_i}^{\eta_{i+1}} \Gamma_I dz = 0. \quad (2.60)$$

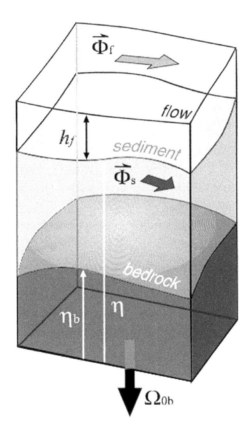

FIGURE 2.25
Definition sketch of the three-layer model used in the generalized Exner equation framework. (From Paola & Voller 2005—reproduced with permission.)

Equation 2.60 is obtained from mass balance consideration at an arbitrarily chosen 'disk' of material I which, at an instant in time, has a volume V fixed by side vertical walls with area A_{side}, two material interfaces with areas Σ_i and Σ_{i+1}, and velocities $\vec{v}_i$ and $\vec{v}_{i+1}$, respectively. $\vec{v} = (u, v, w)$ is the material velocity, and $\vec{\nabla}_H = \left(\dfrac{\partial}{\partial x}, \dfrac{\partial}{\partial y}, 0\right)$ is the horizontal gradient vector. It is important to mention here that outflow of material at the base of the flow layer η implies a positive sign for Ω_{out}, and inflow of material at the top of the flow, $\eta + h_f$, implies a negative sign for Ω_{in}. But what do the terms in Equation 2.60 represent? First, we use Leibniz rule which states that

$$\frac{\partial}{\partial t} \int_{\eta_i}^{\eta_{i+1}} \alpha_I dz = \int_{\eta_i}^{\eta_{i+1}} \frac{\partial \alpha_I}{\partial t} dz + \alpha_I(\eta_{i+1}) \frac{\partial \eta_{i+1}}{\partial t} - \alpha_I(\eta_i) \frac{\partial \eta_i}{\partial t},$$

and then, we identify them one by one for the case of the sediment layer as follows:

- $\int_{\eta_b}^{\eta} \dfrac{\partial \alpha_s}{\partial t} dz =$ change of the sediment density through the sediment column due, for instance, to compaction or inflation,

- $\alpha_s(\eta) \dfrac{\partial \eta}{\partial t} =$ rate of movement of the bed surface η,

- $\alpha(\eta_r) \dfrac{\partial \eta_r}{\partial t} =$ rate of movement of the bedrock interface η_r,

- $\vec{\nabla}_H \vec{\Phi}_s =$ net flow of soil or sediment into or out of the column due to creep in the x–y plane,

- $\Omega_{\text{out}}(\eta_r) =$ net sediment outflow across the sediment–bedrock interface,

- $-\Omega_{\text{in}}(\eta) =$ net sediment inflow across the sediment–flow interface, and

- $-\int_{\eta_r}^{\eta} \Gamma_s dz =$ production or dissolution of particulate mass within the sediment column.

The three conservation equations for the three-layer system can then be combined then into a generalized mass balance equation, which reads as

$$\Omega_{0b} + [\alpha_r - \alpha_s(\eta_r)] \frac{\partial \eta_r}{\partial t} + \int_{\eta_r}^{\eta} \frac{\partial \alpha_s}{\partial t} dz + \vec{\nabla}_H \vec{\Phi}_s - \int_{\eta_r}^{\eta} \Gamma_s dz \quad (2.61)$$

$$+ \alpha_s(\eta) \frac{\partial \eta}{\partial t} + \frac{\partial}{\partial t} \int_{\eta}^{\eta + h_f} \alpha_f dz + \vec{\nabla}_H \vec{\Phi}_f - \Omega_{\text{in}}(\eta + h_f) - \int_{\eta}^{\eta + h_f} \Gamma_f dz = 0.$$

As with all equations, it is good practice to define the terms so we have a good insight into how to modify Equation 2.61 for specific case scenarios. Term 1, Ω_{0b}, is the flux in the sediment–rock boundary, which is positive when we have subsidence, or rock removal, and negative when we have accretion of the rock level or uplift. Term 2, $[\alpha_r - \alpha_s(\eta_r)]\frac{\partial \eta_r}{\partial t}$, is the rate of subsidence or uplift of the sediment–rock interface, times the change in bulk density across the interface. Term 3, $\int_{\eta_r}^{\eta} \frac{\partial \alpha_s}{\partial t} dz$, is the rate of compaction or dilation of the sediment column. Term 4, $\vec{\nabla}_H \vec{\Phi}_s$, is the horizontal divergence of the particle flux within the sediment layer. Term 5, $-\int_{\eta_r}^{\eta} \Gamma_s dz$, is the rate of change of particulate mass in the sediment column, through geochemical processes. Term 6, $\alpha_s(\eta)\frac{\partial \eta}{\partial t}$, is the rate of change of the sediment surface or the water–sediment interface. Term 7, $\frac{\partial}{\partial t}\int_{\eta}^{\eta+h_f} \alpha_f dz$, is the rate of change of sediment mass within the flow. Term 8, $\vec{\nabla}_H \vec{\Phi}_f$, is the horizontal divergence of sediment flux within the flow. Term 9, $-\Omega_{in}(\eta + h_f)$, is the vertical rate of change of mass through the top of the flow layer. Finally, term 10, $-\int_{\eta}^{\eta+h_f} \Gamma_f dz$, is the rate of creation or destruction of sediment mass within the flow. Dissolved phase balance equations have strong parallels with Equation 2.62, as the different terms refer to chemical rather than sediment (or other particulate matter, but we focus here on sediment to avoid confusion) components.

Paola & Voller (2005) discuss a number of examples that can be deduced from the general Equation 2.62, such as the standard Exner equation, the sediment precipitation and sedimentation balance equation, the equation for basin subsidence, and sediment uplift with soil formation. One of the first examples introduced by Paola & Voller (2005) is the standard Exner equation, expressed in the form

$$\alpha_s(\eta)\frac{\partial \eta}{\partial t} + \rho_s \frac{\partial}{\partial t}\int_{\eta}^{\eta+h_f} C_f dz + \vec{\nabla}_H \vec{\Phi}_f = 0, \qquad (2.62)$$

which assumes that changes in the bulk density are due to changes in the volumetric sediment concentration, C_f, in the flow. If sediments are noncohesive, then terms 2 and 3 in Equation 2.62 are expressed in terms of the bed shear stress, as seen in earlier sections. In turn, the shear stress affects the evolution of the topography, through the friction velocity generated by the flow near the bed.

Another case study mentioned by Paola & Voller (2005) is the sediment precipitation and sedimentation balance equation. The short-term Exner balance equation for sediment precipitating in the water involves three terms: changes in bed elevation, particulate build-up, and precipitation or dissolution in the water column, so the balance equation is of the form

$$\alpha_s(\eta)\frac{\partial \eta}{\partial t} + \frac{\partial}{\partial t}\int_{\eta}^{\eta+h_f} \alpha_f dz - \int_{\eta}^{\eta+h_f} \Gamma_f dz = 0. \qquad (2.63)$$

When Γ_f is positive, we have particulate dissolution (i.e., particulate production), and if it is negative, we have sediment precipitation.

The final example is the mass balance equation for fine-grained and cohesive sediment environments, such as the case of mudflats discussed in Section 2.6. In such environments, sediment transport can be thought of as increasingly less reversible, as cohesive effects become dominant. A simple approach is to formulate the erosion and deposition terms separately. Assuming the eroded surface is bedrock, the Exner balance equation for fines,

$$\alpha_r(\eta)\frac{\partial \eta_b}{\partial t} + \frac{\partial}{\partial t}\int_\eta^{\eta+h_f}\alpha_f dz + \vec{\nabla}_H\vec{\Phi}_f - \int_\eta^{\eta+h_f}\Gamma_f dz = 0, \qquad (2.64)$$

involves a (negative) term (term 1) representing erosion from the bed, a temporal or spatial increase in sediment concentration or flux (terms 2 and 3, respectively) representing the particulate fraction of eroded mass, and a dissolution term (term 4) for negative Γ_f—as in Equation 2.63.

2.6   Mudflats and Wetlands

Intertidal flats are characterised by runnels, consisting of small channels that divide a mudflat into ridges. Most of the ebb tidal waters flow seawards through these runnels. Flows on mudflats are extremely shallow flows (Fagherazzi & Mariotti 2012) that cause sediment resuspension in the runnels, as the speeds are comparable to the speeds during ebb and flood velocity maxima. Mudflat runnels tend be a few 100 m long, 0.5–1 m wide, and a few centimetres deep (Whitehouse et al. 2000, Carling et al. 2009). According to Carling et al. (2009), ridges form by accretion through deposition of sand and silt on their surface, while runnels form by incision. Some authors have observed that ridges and runnels at the Humber estuary, UK, are caused by Christie et al. (2000). Runnel incision is favoured by more water being in the runnels than in the ridges during the ebb tide, making runnels more erodible. When runnels are submerged, they may form by internal waves (Dyer 1982) or by helical flows that generate areas with high (low) shear stresses that trigger erosion (deposition), as seen by Williams et al. (2008); such mechanisms may generate the system of runnels and ridges observed in subtidal regions in the Upper Gulf of California near the mouth of the Colorado River, for example (Álvarez, Suárez-Vidal, Mendoza-Borunda & González-Escobar 2009).

Fagherazzi & Mariotti (2012) highlight that mudflat hydrodynamics is very different depending on the presence or absence of runnels. Mudflat velocities can be either flood or ebb dominated when there are no runnels. However, two mechanisms of importance may cause the concentrations to be, on average, flood dominated: the turbid tidal edge (Christie & Dyer 1998) and the channel spillover (Mariotti & Fagherazzi 2012). Bottom shear stresses in runnels need to be modified to account for the interaction of the flow with the runnel boundaries. This is reflected by the introduction of the hydraulic radius computed for the trapezoidal runnel cross-section,

$$R_h = A/P,$$

with A = the cross-sectional area of the flow and P = the wetted perimeter; with the Strickler–Manning coefficient, n, an empirical coefficient of the order of 10^{-2} m$^{-1/3}$, then included in the formulation for the bottom shear stress due to the current (Mariotti & Fagherazzi 2012),

$$\tau_{\text{curr}} = \frac{\rho g n^2 U^2}{R_h^{1/3}}, \tag{2.65}$$

with ρ = water density, g = gravity acceleration, and U = the average velocity of the flow.

The runnel flows studied by Fagherazzi & Mariotti (2012) are very shallow flows, in some cases with water depths under 20 cm. This clearly poses a number of challenges for *in-situ* measurement techniques, which require high specialisation in small scale velocity and SSC measurements. Fagherazzi & Mariotti (2012) measured the water depths every 10 cm with 10 cm blanking, so a minimum depth of 20 cm is required to obtain data. Generally, optical backscatter measurements suggest that the suspended sediments have similar properties as the sediments found on the top surface of the mudflat, which simplifies the field and numerical experiments required to analyse mudflat dynamics. The most important consideration is to determine the distribution of grain sizes of the substrate, and the mud–sand fraction, in order to assess the cohesiveness of the sediment. This can then be included in the Exner equation in order to predict accurately mudflat morphodynamics.

Mudflat dynamics are affected not only by the cohesiveness of the sediments, but in many places also by the presence of vegetation. This vegetation turns the mudflat into a salt marsh, at northern latitudes, or into a mangrove, at latitudes close to the equator. Several experimental studies have been undertaken to evaluate the effects of vegetation on incoming wave energy dissipation. In the case of laboratory work, most experiments had to be performed at small scales and with synthetic salt marshes, built from plastics or other materials. However, new flume infrastructure developments now have dimensions that allow for experiments at full scale to be performed. Such is the case for the study of Möller et al. (2014) in a flume 300 m long, 5 m wide, and 7 m deep, located in Delft. Natural salt marsh vegetation sections were placed in a section of the flume, and full-scale incoming wave and water-level conditions were produced. The experimental set-up is shown in Figure 2.26. These conditions consisted of regular and irregular nonbreaking waves of heights up to 0.9 in 2 m water depth above the vegetated bed. This range included storm conditions. Several soil and vegetation parameters were shown to be statistically equal in the field and in the flume: the soil bulk density, the stem diameter, and stem

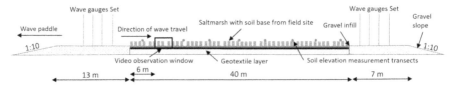

FIGURE 2.26

Salt marsh experiments in the Delta flume by Möller et al. (2014): schematic of experimental set-up.

flexibility. When comparing energy dissipation with and without vegetation, they could attribute 60% of the energy reduction to the vegetation, even in the case of highest waves and water levels. Interestingly, regular waves below and above 0.3 m in height and 3.6 s in period produced a different response in the salt marsh vegetation. With low-energy waves, the plants swayed with the waves throughout the wave cycle. However, for stronger waves and currents, the plants bent to an angle exceeding 50° during the forward wave cycle. This bending preceded stem fracture and resulted in the loss of 31% of the total biomass after 2 days of runs under high incoming wave energy conditions. Once this loss of material occurs, wave energy dissipation by the salt marsh is much reduced. The data produced during these experiments can be used to improve the representation of drag and friction effects in numerical models of wave dissipation and vegetation movement under storm conditions. As Möller et al. (2014) further suggest, the results also support the incorporation of salt marshes into coastal protection schemes, such as the Dutch 'building with nature' approach, which will be discussed in more detail in Chapter 4 but in the context of the Sand Engine.

We have so far reviewed two sample investigations on mudflat and wetland dynamics, both at small to medium spatio-temporal scales. Now we present some research on wetlands in sheltered environments, which are threatened by anthropogenic activities and rising seas (Kirwan & Megonigal 2013). The processes of relevance under such conditions are shown in the schematic of Figure 2.27. The role of human activities is crucial when analysing wetland evolution, because these activities can alter significantly the sediment and water supplied to the system. If this alteration in supply is, for the moment, not taken into account, then the soil elevation simply needs to increase faster than sea-level rise for the wetland vegetation to survive in place (Reed 1995). In order to increase soil elevation, two biological and physical feedback loops, above and below the ground, couple the rate of sea-level rise with the rate of vertical soil accretion. Above ground, sediment deposition is influenced by wetland water depth and inundation periods, with deep wetland areas that inundated more frequently being the ones with largest deposition rates, and with the shallow areas that are inundated less frequently being the ones with less deposition. Deposition is enhanced by the presence of vegetation, which slows down the flow speed and favours sedimentation over sediment

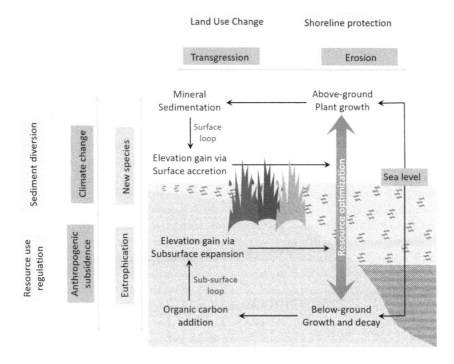

FIGURE 2.27
Schematic of wetland cycles and processes relevant for wetland evolution.
(Modified from Kirwan & Megonigal 2013.)

advection. Below ground, root growth and decay causes subsurface expansion.
Wetland inundation, plant growth, and bed accretion all tend to stabilise most
wetlands; thus there are very few examples of wetland loss driven by sea-
level rise (Kirwan et al. 2010). However, in some systems, inundation does
not stimulate organic matter production any more, and both the inundation
and the reduction in organic matter cause erosion, which leads to wetland
deterioration. Several numerical models have shown that the tidal range and
the sediment available for wetland accretion determine the transition from
stable to unstable wetland. The impacts of sea-level rise, in combination with
anthropogenic activities, can be assessed with numerical models when the
impact of different scenarios needs to be analysed. This is crucial for decision
makers and coastal managers who are interested in the identification and
implementation of the most sustainable wetland management solutions.

Kirwan et al. (2010) provided an analysis of the effects of sea-level rise and
anthropogenic activities on wetlands, based on five different numerical models
(Morris et al. 2002, Temmerman et al. 2003, D'Alpaos et al. 2007, Kirwan &
Murray 2007, Mudd et al. 2009). The models included the ecogeomorphic ability of the wetland to adapt to changes, through nonlinear feedbacks between

processes promoting wetland conservation: inundation, plant growth, primary productivity, and sediment deposition. Morris et al. (2002) showed that a stable equilibrium is obtained when considering two complementary processes: an increase in relative sea-level rise can be compensated by vegetation growth and organic matter production. Morris et al. (2002) considered a salt marsh case study where the dominant vegetation is composed of *Spartina alterniflora*. Because sediment deposition occurs at slower time rates than the interannual anomalies and long period cycles observed in sea-level signals, there is significant variability in annual primary productivity. However, when the bed elevation is larger than what is optimal for primary production, the system is stable despite changes in sea level. When the opposite holds, i.e., the bed elevation is lower than what is optimal for primary production, the system is unstable. Morris et al. (2002) also identified an optimal rate of relative sea-level rise (RSLR), for which the bed-level changes and the depth of the tidal inundation are at equilibrium and plant growth is optimal. Also, when the RSLR is above its optimal value, vegetation growth cannot sustain the necessary bed-level increases that are necessary to maintain the equilibrium. Morris et al. (2002) and Temmerman et al. (2003) showed that the incoming SSCs affect wetland stability, with Temmerman et al. (2003) pointing out that salt marsh erosion occurs with decreases of either the inundation depth or the incoming SSCs. So it is essential to consider both inundation and sediment renourishment sources in wetland morphodynamics modelling.

In a complementary study carried out by D'Alpaos et al. (2007), the authors analysed the effects of sediment transport, vegetation growth, and sea-level rise on the network of flooding and ebbing channels that characterise intertidal flat ontogeny D'Alpaos et al. (2005) and evolving landscapes. D'Alpaos et al. (2007) showed that marsh ecology is a crucial driver of long-term wetland morphological evolution. The authors showed that if the vegetation is dominated by one single species, namely by *Spartina*, then under constant rates of sea-level rise, unvegetated and *Spartina*-dominated environments tend to an equilibrium at or below Mean High Water Level (MHWL) under adequate sediment supply and vegetation growth conditions, but marshes with multiple vegetation species can make a transition upland. By tracking the growth of the intertidal platform, the evolving channel morphology and the vegetation dynamics, D'Alpaos et al. (2007) were able to analyse the mechanisms driving the sedimentation patterns observed in tidal environments, based on different SSCs, relative MSL rise conditions, and absence or presence of vegetation.

The last two models to discuss are those by Kirwan & Murray (2007) and Mudd et al. (2009). As done by D'Alpaos et al. (2007), Kirwan & Murray (2007) analysed landscape evolution and platform changes in wetland environments, through the coupling of geomorphic and ecological models. In addition to other factors mentioned by previous authors, Kirwan & Murray (2007) also considered the creek bank slumping and other slope-driven transport processes that are driven by vegetation biomass. They analysed different sea-level rise scenarios, starting from a steady moderate rise and increasing the rate of sea-level rise. They showed that wetland vegetation can maintain the wetland environment in a metastable equilibrium, even under rapidly rising sea levels, demonstrating the importance of vegetation on wetland habitat conservation. Kirwan & Murray (2007) and Mudd et al. (2009) considered a model that can reproduce wetland platform and channel network evolution. Mudd et al. (2009) considered the above-ground and below-ground organic matter cycles—schematised in Figure 2.27. For the above-ground organic production cycle, they defined the sediment deposition rate as a function of vegetation density, and horizontal distance from a channel, but did not include channel erosion. However, without erosion, the impact of an expanding tidal prism on channel network erosion and expansion, for example, and the subsequent reduction in wetland area (Allen 1997), cannot be taken into account.

We conclude this section with a brief description of the wetland processes included in the model developed by Kirwan & Murray (2007). First, each model iteration is initiated with the high-tide water level, and the topography is partitioned into 5 by 5 m cells. Then, the total draining volume of water per cell for a single ebb flow is computed. The water drains off of the marsh platform, through the channel network, and out of the basin according to flow directions defined by a parametrically represented water surface, without the need for intratidal velocity computations. The volume of water, V, that flows through a given cell is related to high-tide level, s, and the bed elevations, $z(x, y)$, of its watershed, w (Rinaldo et al. 1999):

$$V = \int_w [s - z] \, dx dy. \tag{2.66}$$

The maximum value of $s - z$ is limited by the tidal range, and so cells with elevations that are below low tide do not drain completely. The volume of water is computed over a characteristic timescale of 3 h, to obtain the near-peak water discharge. The discharge must be divided by some water depth to ensure that velocity is higher in shallower areas and lower in deeper areas for a given discharge (Fagherazzi & Furbish 2001). Then the erosion rate is computed in terms of the bed shear stress, $\tau_b = \rho f_c U_*^2 / 8$, as

$$\text{Erosion rate} = \frac{m \left(\tau_b - \tau_c \right)}{\tau_c}, \tag{2.67}$$

where ρ is water density, $f_c = 0.02$ is a friction factor, U_* is the bottom friction speed, $m = 0.0014$ kg/m^2 per second, and $\tau_c = 0.4$ N/m^2 is the critical shear stress, defining the threshold of erosion (Fagherazzi & Furbish 2001). Most erosion occurs over intermittent, near-peak flows, which occur only once over the tidal cycle and have a duration of around 10 min. So the erosion rate is multiplied by 10 min to compute the erosion rate over a single tidal cycle. The model also includes formulations for the deposition rate, the biomass productivity, and the slope-driven sediment transport. Finally, a number of experiments are carried out to model the evolution and dynamic equilibrium of the marsh platform–channel morphology; this dynamic equilibrium is reached when the elevation change everywhere equals the rate of sea-level rise, and the water depth remains constant. This simple model captures several wetland platform and channel network characteristics; for example, the channels become wider downstream. Also, the platform morphology, water depth, and rates of biomass productivity agree well with observations.

3

Modelling Procedures

3.1 Modelling Concepts and Methods for Different Scales and Applications

The fundamental physics of estuaries and coastal environments were introduced in the last chapter. The formulations discussed were obtained through laboratory, field, and remote sensing observations and provide an understanding of the evolution in space and time of sediment concentration fields, seabed heights and estuarine depths and widths, bottom shear stress formulations and sediment fluxes, amongst others. The dynamics of these environments may be analysed using dimensionless parameters: the Reynolds number, the Froude number, the Shields parameter, to name but a few. In the case of wave dynamics, some dimensionless parameters may be rather simple, such as the wave steepness, i.e., the wave height to wavelength ratio, or the relative water depth, i.e., the water depth to wavelength ratio. As explained in Chapter 2, dimensionless parameters are most useful because their size determines important physical characteristics of the system; in particular, in limiting cases, some terms in the equations of motion may be neglected, and importantly, this will lead to simplified equations that can be solved analytically. For example, the relative water depth is an essential parameter in wave dispersion processes, because it determines whether waves are dispersive or nondispersive and whether the wave height, length, and celerity are influenced by water depth (Benassai 2006).

In any modelling effort, it is necessary to inspect the boundary and initial conditions that are most appropriate at the scales of interest. The boundary parameters and variables that need to be defined also depend on the type of boundary. In coastal area models, one needs to consider three types of boundaries. At the ocean or land–ocean boundaries, the incoming hydrodynamic conditions need to be defined, such as waves, tides, storm surges, coastal inlet, or estuarine discharges. At bottom boundaries, the bathymetry (a map of the seabed) or the *topobathy* (a map combining topography and bathymetry), the bottom characteristics and associated model parameters, such as the bottom roughness, need to be considered (Van Dongeren et al. 2013, Kono et al. 2018). And finally, at atmosphere–ocean (either oceanic or estuarine) boundaries, water surface pressures, heat fluxes, and wind shear stresses need to be analysed. In biogeochemical and water quality problems, one may include, additionally, advection-diffusion-reaction-decay equations for sediments, pollutants, and nutrients. In other applications, one may also need to consider chemical transfer equations between the water and the atmosphere. Passive tracers can be analysed following dispersion concepts, thus avoiding the complexities associated with chemical or biological activity; such an approach is valid, for example, for fish larvae dispersion in upwelling regions.

Therefore, generally in morphodynamic models, we will need to define the following forcing parameters (Roelvink & Reniers 2012): the tidal characteristics (amplitude or phase) over a spring–neap cycle; the offshore wave conditions at a representative location or at several locations along a model boundary—significant wave height, period, and direction, as well as the wave energy spectral shape; sea-surface wind speed and direction and sea-level pressure when available; river discharges (and constituent characteristics, when relevant); and surge level during extreme event conditions. As Roelvink & Reniers (2011) point out, care must be taken with correlated parameters. For example, the surge level is often correlated (with lots of scatter) to the wind and (for a given direction) to the waves; the locally generated wave direction is correlated with the wind direction, especially under strong winds; the locally generated wave heights and wind speeds are correlated, for a given direction; and the locally generated wave period is correlated with the wave height. The tidal boundary conditions are uncorrelated with other forcings, although tides may be modified by surges, winds, and waves.

3.1.1 Mesh Generation

3.1.1.1 Mesh Design in the Horizontal Coordinates

Designing a good quality mesh, whether it is a spherical or a topologically Cartesian one, is an essential consideration in the model design process, in particular around the area of interest. Several modelling suites have their mesh generator, which facilitates and speeds up the process. In this section, we will use numerical tide and storm surge modelling examples. Numerical storm surge modelling and prediction intensified in the countries bordering the North Sea after the great storm surge of 31 January to 2 February 1953, which caused severe coastal damage in the Netherlands, the UK, and Germany. Due to this storm, coastal management policies changed significantly in these countries. Extreme event modelling is obviously important for a number of reasons, as pointed out by Flather (1984). Firstly, such events may cause severe damage and loss of lives around the world, and therefore any forecasting model needs to be able to reproduce the impacts of such events. Secondly, numerical models give spatio-temporal information on storm surge evolution that is generally impossible to obtain by other means. Finally, the numerical model predictions of surge and tidal water levels can be used for coastal defence design.

We will first discuss the grid generation options on the horizontal coordinates or the latitude and longitude coordinates in large geographical domains. Along these two coordinates, the models may use regular Cartesian grids, regular curvilinear grids, unstructured meshes, or a combination of those. Regular Cartesian grids were common in early modelling efforts, for example, in numerical storm surge modelling by Hansen (1956, cited by Flather 1984) and (Prandle & Wolf 1978). Due to their relevance and simplicity, they are still used in current applications. Cartesian grids may be have uniform, nonuniform ('plaid') or adaptive grid spacing (Van Hooft et al. 2018) and may be locally refined, as in quadtree models (Popinet & Rickard 2007). Two methods are available for local grid refinement, namely, domain decomposition (Funaro et al. 1988) and nesting methods (Cailleau et al. 2008). They both are implemented in a number of ocean and coastal hydrodynamics models, such as in the Dutch codes TRISULA (currently known as Delft3D-FLOW) or SIMONA, both for Cartesian and for regular curvilinear grids (de Goede et al. 1995). Figure 3.1 shows a close-up of a Cartesian grid (Figure 3.1a) and a curvilinear grid (Figure 3.1b), for a coastal region in the Upper Gulf of California, around San Jorge Island, which is in fact an archipelago of smaller islands. The largest island is very long, with a length of around one kilometre, and very narrow, with a width of around 185 m at its widest location. Its northern beach (the only beach), used by the seal colony, is geographically located at $(31°00'59.47''N, 113°14'42.22'' \text{ W})$.

Curvilinear grids evolved from Cartesian meshes as a more advanced level of complexity (Amsden & Hirt 1973), even if curvilinear grids are topologically equivalent to Cartesian grids. A curvilinear grid has important advantages in comparison to a Cartesian grid, as by design the grid can adjust to offshore structures or follow coastlines or isobaths. Like Cartesian grids, curvilinear grids are structured meshes, except they may not be orthogonal. However, some orthogonality conditions usually are necessary, for code robustness

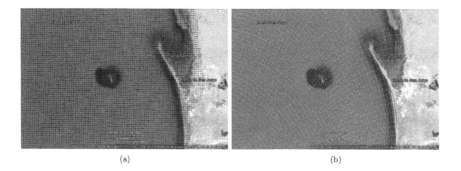

(a)   (b)

FIGURE 3.1
Illustration of a structured regular grid with a resolution of 400 m × 400 m and a curvilinear grid with varying resolution, using grid close-ups for numerical coastal models of a region in the Upper Gulf of California (Mexico). (a) Close-up of a structured Cartesian grid and (b) close-up of a structured curvilinear grid.

and stability. In Delft3D for example, the curvilinear coordinates are denoted as (ξ, η), to differentiate from the Cartesian (x, y) coordinates. Orthogonality and smoothness determine the quality of a curvilinear grid. The angle, φ, or the cosine of the angle, $\cos(\varphi)$, between the ξ and η direction is a measure of grid orthogonality, and the condition $\cos(\varphi) < 0.02$ ensures good grid quality. A measure of grid smoothness is the aspect ratio of the grid cells, i.e., the ratio of the grid cell dimension in the ξ and η direction, and it should be between 1 and 2 to ensure aspect ratio smoothness, except when the flow is predominantly along one of the grid lines. Another measure of grid smoothness is the ratio between neighbouring cell dimensions, which should be <1.2 in the area of interest, and <1.4 further away (Deltares 2018). The grid shown in Figure 31b is a close-up of a curvilinear grid covering the same study site. The orthogonality condition on the grid is satisfied, and the smoothness condition is also satisfied, even if the aspect ratio for cells far from the coast is close to 2. The grid cell size depends on the patterns that need to be resolved, but for example, to resolve some circulation patterns, the grid cell size should be at most one-tenth the size of the circulation. To resolve a certain bathymetric or hydrodynamic feature, the grid cell size should be at most one-fifth the size of that feature. Moreover, in areas where flooding and drying is relevant, such as in macrotidal, shallow environments, the flooding and drying schemes are more accurate with fine grid resolutions.

Other mesh generation options consist of unstructured meshes, such as finite elements. The newest Delft3D model for example, Delft-FM, is based on unstructured grids. Unstructured grids consist of map tessellations, where closely fitted shapes define the grid cells within the mesh. For coastal modelling applications, these unstructured grids may be generated with triangles, tetrahedra, or other polygonal shapes, depending on the complexity of the domain coasts and bathymetries. Figure 3.2 shows an example of an unstructured grid.

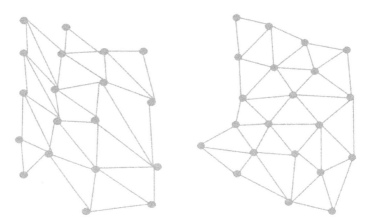

FIGURE 3.2
Illustration of an unstructured grid.

The grid on the left is of poor quality: it is significantly skewed and has poor smoothness and orthogonality properties. The grid on the right, however, is of good quality.

The choice of grid has significant implications for the speed of convergence, solution accuracy, and CPU requirements. In structured grids, the mesh directions are easy to identify by associating a curvilinear coordinate system and finding the nearest neighbours of any node point. Also, many algorithms for discretisation of the equations are available, and they can be implemented in an efficient manner. However, they are difficult to implement in regions with complex geometries, where block structured grids need to be implemented (these blocks are naturally parallelised). In contrast, unstructured grids have irregular patterns, so mesh directions need to be carefully defined, and they have large storage requirements; but they offer the possibility of great adaptability. The mesh generator also needs to be compatible with the solver. In the case of Delft3D, the shallow-water equations are solved using finite differences, and for these discretisation processes, it is important that the grid lines correspond to lines with constant coordinate values. Finite element and finite volume codes use unstructured grids, as they are written only for meshes with certain shapes such a triangles of tetrahedrons. One condition in mesh generation is that the nodes of adjacent shapes are the same and that the element angles satisfy the orthogonality condition; otherwise, accuracy in the solution is lost, the discretised matrices may be ill conditioned, or the code may fail to converge. Also, as for structured grids, the resolution of the unstructured grid needs to be high enough to resolve the flow patterns under study. For example, meshes need to be finer in boundary layers, near coastal complex features, such as rivers or estuaries, or within vortex cores, among others. Finally, meshes that can be refined quickly and easily may be preferable. Triangles and tetrahedrons, for example, may be generated with simple mesh-generating algorithms, such as that by Persson & Strang (2004). More details on mesh-generation techniques may be consulted in Frey & George (2008). Several off-the-shelf models include a mesh-generation software. Figure 3.3 shows a sigma-layer, curvilinear grid on the left, and a z-layer grid on the right. Sigma layers follow lines of constant pressure and, thus, adjust to the bottom and the surface boundaries. Z layers are useful when the surface layer properties in deep water regions need to be resolved with good accuracy.

3.1.2   Model Topography and Bathymetry

In applied projects and in local or regional modelling case studies, a detailed bathymetry obtained with high-resolution instrumentation is ideal. As discussed in Chapter 2, highly detailed bathymetries can be generated with echosound or sonar technologies. At the other end of the spectrum, we have low-resolution bathymetries, which can be determined from satellite imagery; the bathymetry is inferred from sea-surface deformations. Here we present a few examples of global-, regional-, and local-scale bathymetries.

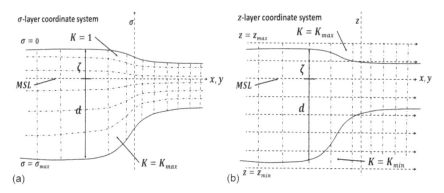

FIGURE 3.3
Illustration of σ and z layer coordinates.

3.1.2.1 ETOPO1 Global Relief

ETOPO1 (doi:10.7289/V5C8276M) is a global-scale bathymetry with a 1 arc-minute resolution; it incorporates both ocean bathymetry and land topography. This topobathy was released in 2008 by the National Geophysical Data Center (NGDC), part of the National Oceanic and Atmospheric Administration (NOAA). ETOPO1 is freely available from its official website. There are two versions which differ at the poles, the *ETOPO1 Ice Surface* showing the ice sheet topographies and the *ETOPO1 Bedrock* showing the bedrock under the ice sheets. This dataset is horizontally georeferenced to the World Geodetic System 1984 (WGS84) and vertically to the mean sea level (MSL). This bathymetry was obtained from various sources, shifted to the same horizontal and vertical datum, and interpolated at the same 3D grid nodes. These sources include, for example, the NGDC, the Antarctic Digital Database (ADD), the European Ice Sheet Modeling Initiative (EISMINT), the Scientific Committee on Antarctic Research (SCAR), the Japan Oceanographic Data Center (JODC), the Caspian Environment Programme (CEP), the Mediterranean Science Commission (CIESM), the National Aeronautics and Space Administration (NASA), the National Snow and Ice Data Center (NSIDC), the Scripps Institute of Oceanography (SIO), and the Leibniz Institute for Baltic Sea Research (LIBSR). Maps of global relief may be downloaded directly from the ETOPO1 website, and data may be retrieved in various formats, including netCDF, xyz, or georeferenced TIFFs.

3.1.2.2 Bathymetries with 30 Arc-second Grids: GEBCO2014 and SRTM30_PLUS

As for the ETOPO1 product, the General Bathymetric Chart of the Oceans (GEBCO) combines information from around the world in a single source, which may have a resolution of 1 arc-minute or 30 arc-seconds, depending of the GEBCO version. However, the one arc-minute GEBCO chart was last updated in 2008, and there are no plans of updating it again. The 30 arc-second chart was last updated in 2014 and is the one currently used in many regional models. In the latest version, GEBCO2014, thirty-three percent of the GEBCO2008 bathymetry was updated by Pauline et al. (2015). However, a great disadvantage of GEBCO products is that they assume that the depths are referred to MSL when assimilating data products, which is not always the case, and therefore at times, the GEBCO data close to the coast may not be very reliable. The GEBCO working group is evaluating how to resolve this issue in future releases (as read in www.gebco.net/data_and_products/gridded_bathymetry_data/, on the 11th of July 2018). The 2014 bathymetric chart is described and discussed in (Pauline et al. 2015). The GEBCO project started in 1903 with funding from the Monaco monarchy, with the objective of mapping the world's oceans, using ship soundings. By 1973, GEBCO had transformed into an international program, with funding from the International Hydrographic Organization (IHO) and the Intergovernmental Oceanographic Commission (IOC) of the United Nations. As for all bathymetric chart products, GEBCO is a joint effort by bathymetric data providers and grid developers, combining several international bathymetric charts into one global product. Figure 3.4a shows the bathymetry of the Gulf of California from the GEBCO dataset.

Figure 3.4b shows the same area produced with data from the Global Shuttle Radar Topography dataset SRTM30_PLUS, another bathymetric data product with a 30 arc-second resolution. The Space Radar Topography Mission (SRTM) is discussed in more detail in the following section, but the SRTM missions have focused mostly on topographic analyses of the Earth surface, and therefore the SRTM is likely to use the same data sources as GEBCO for the generation of the bathymetric estimates reported in the SRTM literature

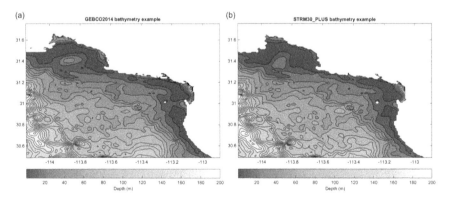

FIGURE 3.4
Illustration of a GEBCO2014 and a SRTM30_PLUS bathymetry for the Upper Gulf of California. (a) GEBCO2014 bathymetry and (b) SRTM30_PLUS bathymetry.

and databases. This is confirmed by the fact that both datasets generate similar bathymetries for the North Eastern side of the Gulf of California, as illustrated in Figure 3.4. Also, it is unlikely that the bathymetry has been quality controlled in detail. In fact, the SRTM documentation only indicates that the land topography is fused with measured and estimated seafloor topography, so it is the responsibility of the user to check the validity of the bathymetric data, in particular close to the coast, where it has been shown that GEBCO has inaccuracies in the bathymetric estimates, with important consequences for coastal studies (Magar et al. 2018). Despite this, the SRTM data is an excellent source of topographic data, and having access to such high-quality data is crucial, for instance, for studies on air–sea and land–sea interactions and coastal inundation under hurricane forcing conditions (Kono et al. 2019).

3.1.3 More Detailed Bathymetries with Satellite Technologies

The Global Shuttle Radar Topography mission topography and bathymetry data SRTM15_PLUS, corresponds to a resolution of 15 arc-seconds. The data is available with ftp protocol at this website. The land topography is based on SRTM, ASTER, and CryoSat-2 ice sheet topography for the ice-covered areas. The land topography in SRTM15_PLUS was completely revised for this version of SRTM, but only the SRTM30_PLUS is publicly available. The bathymetric data includes 494 million seabed soundings, weaved within bathymetric predictions from a gravity model obtained with the Jason-1 and the CryoSat-2 missions. This is the bathymetry used by ©Google Earth. In contrast to the land topography, the only difference (aside to the interpolation to a finer grid) between SRTM15_PLUS and SRTM30_PLUS data is a correction on the borders of the swath trajectories in the SRTM15_PLUS version, as explained in ftp://topex.ucsd.edu/pub/srtm15_plus/README.V1.txt. There are 3 arc-second and 1 arc-second versions of the SRTM mission, which can be found at https://lta.cr.usgs.gov/SRTM1Arc. But the fine resolution is limited, again, to land topography.

Current satellite imagery is so detailed that it is used not only for topographic studies, but also for identifying vegetation coverage. Since 1999, land topography can be evaluated using remote sensing techniques at resolutions higher than 100 m, from some of the following missions:

- NASA LandSat 1–9 missions: Multispectral remote sensing satellite missions, originally called *Earth Resources Technology Satellite* (renamed in 1975). The first LandSat mission (LandSat-1) was launched on 23 July 1972. The first three missions, with at least one operational satellite at all times, had a sampling multispectral scanner with a resolution of ~80 m and a Return Beam Vidicon (RBV) camera system with a resolution of ~40 m. In LandSat-4, launched on 16 July 1982, and in LandSat-5, launched 1 March 1984, the RBV was replaced with a thematic mapper (TM) with a resolution of ~30 m on the reflective bands or ~120 m on the thermal band. LandSat-6 did not reach orbit. LandSat-7, launched on 15 April 1999, and still operational to date (contrary to the previous missions), has a single sensor, an Enhanced Thematic Mapper Plus (ETM+), with resolutions of 15 m on the panchromatic (PAN) bands, 30 m on the reflective bands, and 60 m on the thermal bands. LandSat-8, launched in 11 February 2013, and still operational, has two sensors, an Operational Land Imager (OLI) and a Thermal Infra-red Sensor (TIRS); the OLI has similar resolution as the ETM+ but no thermal bands; the TIRS has two bands, both with a 100 m resolution. LandSat-9 will be launched in December 2020.

- European Space Agency (ESA) Sentinel 1–5 missions: The Sentinel project started in April 2014 with the launching of the Sentinel-1 satellite. The first three missions focused on land and sea monitoring and data acquisition, while the missions 4, 5, and 5P were dedicated to air quality monitoring. The Sentinel missions are collecting data for the Copernicus programme. Further details of all missions may be consulted at www.esa.int/Our_Activities/Observing_the_Earth/Copernicus/ Overview4. Here we provide a brief summary of the characteristics of the first three missions: Sentinel-1 is a polar-orbiting radar imaging mission for land and ocean services; Sentinel-2 is a polar-orbiting, multispectral high-resolution imaging mission with 13 spectral channels at 10, 20, and 60 m spatial resolutions, providing imagery of vegetation, soil and water cover, inland waterways and coastal areas; Sentinel-3 is a multiinstrument mission for sea surface topography, sea surface and land surface temperature, ocean colour and land colour measurements with high-end accuracy and reliability. This mission provides data for ocean forecasting systems, as well as environmental and climate monitoring.

- Advanced Spaceborne Thermal Emission and Reflection Radiometer (ASTER) Missions: ASTER is a Japanese sensor on board of NASA's Terra satellite, launched, in 1999. ASTER data are used to create detailed maps of land surface temperature, reflectance, and elevation. On 1 April 2016, the entire dataset was opened to the public; it may be downloaded free of charge from https://asterweb.jpl.nasa.gov/data.asp.

Data at higher resolutions of <10 m can be collected with the latest satellite technologies, such as GeoEye-1, now DigitalGlobe (1.84 m multispectral resolution), QuickBird, WorldView 1–4 (~0.5 m), IKONOS (0.82 m panchromatic to 4 m multispectral), Pléiades (2 m multispectral), KOMPSAT (0.55–5.5 m at nadir), TripleSat (0.8 panchromatic to 4 m multispectral), SPOT 1–7 (~6 m multispectral), RapidEye (5 m multispectral), or PlanetScope/Dove (2.7–4.9 m). These and other satellite missions are described at www.satimagingcorp.com/satellite-sensors/. The main difficulty at such high resolutions is to find available adequate data and coverage for the case study of interest. However, these high resolution images are particularly useful for bathymetric data in shallow waters, which tend to exhibit rapid changes, both in time and space. One of the challenges for the past 10 years has been to obtain bathymetric information for shallow waters with satellite sensors, because of the uncertainty in the estimated water depths due, for example, to vegetation coverage or to turbidity levels. With ground truthing, some of the uncertainties can be corrected. In fact, ground truthing helps train the bathymetry analysis algorithms Ohlendorf et al. (2011). For instance, the bathymetry for Puerto Morelos shown in Figure 3.5 was calibrated with *in-situ* beach profile measurements (Cerdeira-Estrada et al. 2012), and at these locations, the WorldView-2 agrees almost perfectly with the beach profiles determined from beach GPS surveys. Having such large coastal bathymetric coverage may lead to significant advances in coastal science, because of the large extent and the density of coverage of Satellite data. Echosound sensors may have similar resolution, or even higher, but the areas that can be covered

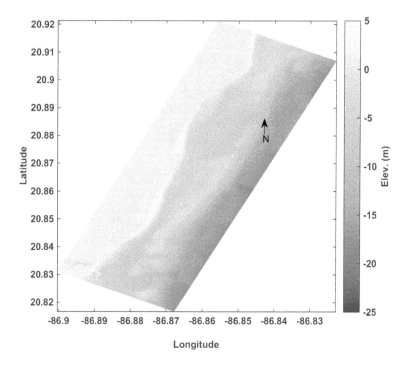

FIGURE 3.5
High-resolution bathymetry: Puerto Morelos Reef. The length of the beach (on a straight line) is of around 12 km. (Courtesy of CONABIO.)

are, in general, smaller than those covered by satellite sensors, and their use is costly and time consuming.

In summary, ocean and coastal modelling has relied on remote sensing methods for many decades. However, only for the last decade or so, when the sub-10 m resolution satellite missions were launched, has it been possible to determine bathymetric and benthic vegetation coverage characteristics from Digital Globe ventures. However, ground truthing and local echosound measurements are still necessary under some circumstances, especially in small areas where it is essential to have data at regular, short intervals, in order to assess bathymetric changes. When one has several bathymetric data available, it is important to test their accuracy, for example, by making difference maps; checking that the bathymetry in areas where volume changes need to be determined has sufficient resolution; studying profile changes and beach behaviour over different cross-sections; making animations of beach evolution; and analysing changes and migration rates of several morphological units, such as ripples, sandbars, sandbanks, dunes, or shorelines. If the bathymetry is to be used in a numerical model, it needs to be interpolated to a grid that is common for all the boundary conditions and forcings. The interpolation can be performed with grid cell averaging if there are many points per grid cell or with internal diffusion if there are very few or no points in the grid cells. Also, it is important to check that the bathymetry is behaving well by analysing important bathymetric cross-sections.

3.1.4   Tidal Forcing: Modelling Approaches

Tides may be introduced as a forcing at the open boundaries of numerical models by determining the amplitudes and phases of the principal tidal components. There are two pieces of software commonly used by coastal engineers and geoscientists, namely T_TIDE (Pawlowicz et al. 2002) or UTIDE (Codiga 2011). The general equation of the reconstructed water elevation ξ_{ast} (*ast* for *astronomical*), based on N tidal components with amplitude A_i, frequency ω_i, and phase ϕ_i, is (Godin 1972)

$$\xi_{\mathrm{ast}}(t) = \sum_{i=1}^{N} A_i \cos\left(\omega_i t + \phi_i\right). \tag{3.1}$$

The frequency ω_i of each of the tidal components is a constant, but the amplitude and the phase vary spatially. In most modelling efforts, researchers use at most 13 tidal components, determined from the TOPEX–POSEIDON (TPXO) global tidal model (Egbert et al. 1994, Egbert & Erofeeva 2002) (see http://volkov.oce.orst.edu/tides/global.html). TPXO is based on the Laplace Tidal Equation with satellite data assimilation from the TOPEX–POSEIDON missions. The first versions of TPXO provide 13 tidal constituents, including the following eight primary: M_2, S_2, N_2, K_2, K_1, O_1, P_1, and Q_1; two long-period: M_f and M_m; and three nonlinear: M_4, MS_4, and MN_4. TPXO9 includes also the constituents $2N_2$ and S_1. The model versions are improved by improving the bathymetry and by assimilating more satellite data. Thus, in the latest TPXO version, there are 15 tidal components.

A well-established methodology concerning tidal level analysis in regional, morphodynamic models is presented by Roelvink & Reniers (2012). It applies to all sites that are semidiurnal, or predominantly semidiurnal, and comprising the following steps: choose a representative month having an average neap/spring tidal amplitude ratio, and run the tidal model for that month, with time series outputs at representative locations; estimate the average transport over 59 tidal cycles and the average transport over two consecutive tides starting at each flow reversal; select the period for which the average transport over two tidal cycles agrees best with the monthly averaged transport; carry out a Fourier analysis over the two selected tidal periods for each of the boundary support points; and remove the diurnal and the odd tidal components from the time series. The resulting signal corresponds to a single representative tide, the so-called *morphological tide*, resulting in a reliable description of the net sediment transports (Van Duin et al. 2004). Doing the Fourier analyses over two consecutive tides avoids the problem of contamination of the mean by the diurnal tide. If simulations are short term, then this representative tide is a very good option for the water level forcing at the boundary.

If the simulations required cover periods of more than 1 month, then the model can, alternatively, be forced with TPXO data or with several tidal components (15 say), and several short-period models can be set up, each starting and finishing at times that cover, without overlaps, the time series being simulated. The results are then added together (Elias, Personal communication). This method permits long-term simulations to be run much faster, with results that are still reasonable. Another possible option to run long-term simulations is to use domain decomposition to create subdomains which can be run in different computer threads. Then the simulations are performed, in principle, faster compared to models based on only one domain. Therefore, to synthesise the conceptual idea behind these techniques, in one case, we are doing a temporal schematisation and, in another, a spatial (domain) decomposition. One of the difficulties with time- and space-domain decompositions is the passing of information between domains. Passing of information is also a stumbling block with models forced with more than just tides, winds, or waves and with models coupling sediment transport and morphodynamics modules. Such models require a significant calibration effort, which may involve tens of short-period tests (e.g., over two tidal cycles) of the model.

3.1.5   Wave Forcing: Modelling Approaches

Although wave forcings are, in fact, linked to meteorological forcings, they are such an essential component of the dynamics in shallow open coastal regions that we will dedicate Section 3.2 to analyse different models that may be used in wave-driven morphodynamic modelling. Here we will only discuss aspects that have relevance at different scales and models with different levels of detail. There are two sorts of wave models: phase-resolving models and phase-averaged models, as discussed in Section 3.2. In either of those models, wave climate schematisations may be possible for long-period modelling studies.

One of the first assumptions in forcing schematisations is that chronology does not play any role. The schematisations are based on initial transports and bottom changes. The objective of the schematisation is to reduce the number of input conditions that have to be simulated, together with a morphodynamic updating technique. For the case of the wave conditions, the parameters involved are the wave height, period, and direction, as well as the spectral shape at representative locations. It is important to take into account the correlations between wave period and wave height, as well as those between wind wave direction and wind direction, and between wind wave height and wind speed. The tidal elevation is modulated by the wave climate, through refraction for example, but generally the tidal schematisation is performed separately from the waves.

As an illustrative example, 1 year of data has been extracted from NOAA's deep water buoy station no. 46086, in water of depth 1,829 m, maintained by the National Data Buoy Center, located at $32°29'27'N$, $118°2'5'$ W. The buoy records significant wave height, dominant wave period and average wave period, mean wave direction, as well as wind direction and wind speed (see www.ndbc.noaa.gov/ for further details). The wave rose shown in Figure 3.6 is constructed by dividing the data into 8 wave height bins and 12 direction bins. The direction bins show the total occurrence of waves in that direction, as well as the occurrence of each of the eight wave height ranges in each direction. The wave rose shows that most of the waves travel in the south–southeast (SSE) direction, and more than 40% of the waves in that direction have wave heights of 2.4 m or less.

The wave rose is an important tool in identifying the dominant significant wave height and direction of propagation, and once the dominant significant wave height has been defined, the peak period needs to be determined as well. Complex wave models make use of multidirectional wave spectra, but in most cases, the domains are rather small, and focusing on the waves travelling in a dominant direction tends to be sufficient.

3.1.6   Meteorological Forcings

In order to incorporate meteorological effects and wave set-up in the water elevation time series, it is necessary to include wind stresses, surface pressure differences, and wave dynamics at the sea surface. The wave data can be extracted from NOAA boys, from global wave models such as WAVEWATCH III, or from local *in-situ* measurements obtained from Nortek's acoustic wave and current profilers (AWACs), an acoustic Doppler current profiler (ADCP)

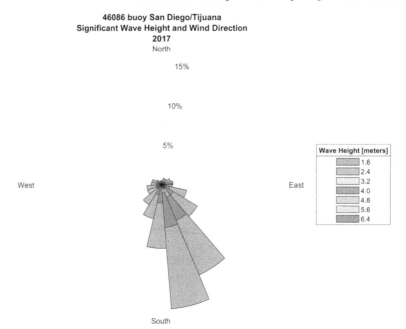

FIGURE 3.6

Illustration of a wave rose diagram using data from the National Data Buoy Centre (NDBC) buoy no. 46086 in San Clemente Basin, off the Californian Coast.

with integrated wave sampling technology, or any ADCP with wave sampling capacity. Modelling waves is important in open beach dynamics, which will be discussed in Section 3.2. However, estuaries and coastal lagoons will only be affected very close to the estuaries' mouths or the lagoons' inlets by incoming ocean waves. In all coastal waters, winds will generate wind-driven currents as well as waves, and wind-driven set-up, and their effects will be more important for water bodies with a very large area, for example, Great Lakes or Marginal Seas, such as the Gulf of California. Wind forcings are also important in small estuaries and coastal lagoons, but since the fetch will generally be smaller in those cases, their effects will be smaller too. There are many wind speed and sea-surface pressure databases, such as the North American Regional Reanalysis (NARR), the North American Monsoon (NAM) database from NOAA's North American Monsoon—Climate Prediction Center, or the Climate Forecast System Reanalysis (CFSR). Wind speeds may also be obtained from satellite data, such as AVISO, QuickSCAT, or Synthetic Aperture Radar (SAR) images. Wind speeds obtained from SAR both near the coast and in the ocean have very good accuracy, even close to complex terrain (Fisher et al. 2008). In fact, SAR yields backscatter images with higher resolution than those obtained from scatterometers. For local studies, however, it is best to identify a local meteorological station with wind speed and direction time series of sufficient length or install one when necessary. Wind forcings for global and regional modelling can also be obtained from numerical weather prediction model reanalyses, such as those provided by the Fleet Numerical Meteorology and Oceanography Center (FNMOC), the National Centers for Environmental Prediction (NCEPs), or the European Centre for Medium-Range Weather Forecasts (ECMWF). Generally, global ocean models are forced with wind fields with grid resolutions of the order of $1°$–$1.5°$, which is equivalent to a resolution of 110–160 kms. But coastal-scale models need winds to be resolved at much higher resolutions, on the order of kms; otherwise the effect of the wind on the coastal ocean circulation or on the sea-surface temperature variability is not well reproduced. Even a 25km resolution can be too coarse for some applications. For instance, such resolution may be justified for a countrywide assessment of wind energy resources (Gross & Magar 2015), but near the coast or for regional or local resource assessments, a finer resolution is advisable, as adopted by Hahmann et al. (2014). At local scales, complex areas with different coast orientations will require some special treatment of the forcing conditions during the morphodynamic model set-up (Roelvink & Reniers 2012). First of all, in order to set up the model with simplified forcing conditions, it is necessary to define a *target sedimentation—erosion pattern*. The target pattern is determined using the following procedure:

- The tide-averaged sedimentation–erosion pattern is computed first using a flow, wave, wind, and morphology model that runs over one tidal cycle. This initial model can have a coarse resolution, if necessary.

- This model is run for each wave, wind, and tide conditions.

- The *target* sedimentation–erosion pattern is the weighted average over all patterns, taking into account the probability of occurrence or 'weight' of each condition.

- The computation is then simplified by finding a set of reduced conditions and weight factors that produce as well the target sedimentation–erosion pattern.

In some cases, regional coastal models require wind data (specifically, wind shear stresses, either an input or computed within the model from wind speed data) and sea-surface pressure. For models with large domains, the density gradients need to be considered as well, as the main driving forces of ocean currents. The sea-surface temperature is also of relevance, as the air–sea temperature difference will drive the heat fluxes between the ocean and the atmosphere. Finally, precipitation and evaporation are also part of the atmospheric forcing on the surface of the ocean. Generally, the wind stress is proportional to the square of the wind speed, with a drag coefficient being the constant of proportionality. As part of the meteorological forcings, one may consider heat fluxes between the ocean and the atmosphere. Although heat fluxes may also be caused by underwater volcanoes or other geothermal activity at the seafloor, once the basics of ocean–atmosphere heat fluxes are understood, it is relatively easy to define heat fluxes between the Earth's crust and the ocean. Heat may be considered as a passive tracer being transported by currents and stirred by waves and by turbulence. Usually only the advective heat flux, Q_{ADV}, is considered. Heat loss or gain takes place at the domain boundaries: at the ocean surface, at the seafloor, at estuaries or rivers, and at the boundaries that are connected with the open ocean. The surface heat flux, Q_{SUR}, has four components as follows:

- sensible heat flux, Q_{SH}—heat transport by conduction
- latent heat flux Q_{LH}—heat transport by evaporation or condensation
- long-wave radiation Q_{LW}—heat transport by back radiation
- short-wave radiation Q_{SW}—heat inflow by solar radiation at the sea surface

with $Q_{\mathrm{SUR}} = Q_{\mathrm{SH}} + Q_{\mathrm{LH}} + Q_{\mathrm{LW}} + Q_{\mathrm{SW}}$, and $Q_{\mathrm{NET}} = Q_{\mathrm{SUR}} + Q_{\mathrm{ADV}}$. The ocean–atmosphere net heat flux may be deduced from numerical weather prediction models, from *in-situ* or remote sensing observations, or both. Heat fluxes based on observations are determined through bulk formulae. For example, the sensitive and the latent heat fluxes,

$$Q_{\mathrm{SH}} = \rho C_P C_S U_{10}(t_s - t_a), \tag{3.2}$$
$$Q_{\mathrm{LH}} = \rho L_E C_L U_{10}(q_s - q_a), \tag{3.3}$$

depend on C_P, the specific heat capacity of air; C_S, the sensible heat transfer coefficient; L_E, the latent heat of evaporation; U_{10}, the wind speed at 10 m above the sea surface; t_s and t_a, the sea and air temperatures, respectively; and q_s and q_a, the specific humidity at the sea surface interface and at 10 m above the sea level, respectively. Alternatively, some models such as the MITgcm, include heat flux computation packages that determine the fluxes from sea-surface temperature fluctuations (see http://mitgcm.org/pelican/online_documents/node216.html for more details). Each of the four components of the surface heat flux have different origin and mean annual value ranges. The annual mean Q_{SH} varies between -42 and -2 W/m^2, and its major contributors are the wind speed and the temperature difference between the ocean and the atmosphere. High wind speeds and large temperature differences cause high Q_{SH}; this is similar to the *wind chill factor* on land. The annual mean of Q_{LH} varies between -130 and -10 W/m^2, and its major contributors are wind speed and the relative humidity of the air above the ocean. The annual mean of Q_{LW} varies between -60 and -30 W/m^2, and its major contributors are heat absorption by clouds and by water vapour content, with a minor contribution from black body radiation and sea ice. The annual mean of Q_{SW} varies between $+30$ and $+260$ W/m^2, and its major contributors are the solar inclination, clouds, and day length; some minor contributors include radiation, reflectivity of the ocean surface, and aerosols and absorbing gases (Monger & Pershing 2005). Salinity changes occur through precipitation, evaporation, and river run-off. Salinity gradients caused by precipitation can extend for several kilometres and persist for several hours under low wind conditions (Soloviev & Lukas 1996, Boutin & Martin 2006). Salinity gradients due to precipitation are called negative salinity anomalies because the salinity decreases with decreasing depth, while positive salinity anomalies occur when the salinity increases with decreasing depth. Temperature and salinity gradients cause changes in density and are, thus, linked to ocean stratification. Density gradients are generally stronger where either salinity or temperature gradients are dominant influences, although there are other factors affecting density gradients, including advection and mixing, both in the horizontal and in the vertical (Hosegood et al. 2008).

3.2 Modelling Open Beach Dynamics

Open beach dynamics, in contrast with the dynamics of estuaries and coastal lagoon, is strongly influenced by gravity and infragravity waves. Following the geographical definition of a beach discussed in Section section 1.1 (see Figure 1.3), it may be inferred that beach dynamics modelling focuses on the coastal zone between the depth of closure and the coastal dune (or cliff) system on the shoreface. This modelling may focus on the beach extent or, as is more frequently the case, on the dynamics of part of the beachface, that is, either the surf zone or the swash zone, or the submerged sandbar dynamics, or the coastal dune dynamics, for example. In some cases, some natural or man-made structures, such as coral reefs or breakwaters, cause fluid—structure–sediment interactions. These interactions are complex in nature, and models need to be chosen with care. Mostly, models differ in the wave modelling formulations and may be divided into spectral wave models or phase-resolving models.

3.2.1 Morphodynamic Models with Spectral Wave Formulations

Spectral wave models are based on the wave action conservation equation, or WAE, which describes the evolution of the wave spectrum, in terms of the action spectral densities. The most general formulation of the wave action is given in Andrews & Mcintyre (1978). Here we assume a linear and irrotational wave theory framework, where such wave action spectral density may be defined as

$$A(k, \theta) = \frac{E_w(k, \theta)}{\sigma} = \frac{E_w(f, \theta)}{2\pi c_g(f, h)\sigma}, \tag{3.4}$$

with E_w the wave energy density and σ the relative radian frequency. For linear waves, the wave group velocity c_g only depends on the frequency, f, and the water depth, H, as stated in Equation 3.4. The wave action conservation equation in Cartesian coordinates satisfied by A in slowly varying depths and flows may be written as

$$\frac{\partial A}{\partial t} + \nabla_x \left[(\vec{c}_g(k,\theta) + \vec{u}) \, A \right] + \frac{\partial}{\partial k} \dot{k} A + \frac{\partial}{\partial \theta} \dot{\theta} A = \frac{S_{tot}(k,\theta)}{\sigma}, \qquad (3.5)$$

where S_{tot} is the total energy source term, with all sources and sinks contributing to the changes in wave energy except for the adiabatic energy exchange due to varying currents (Phillips 1977). This formation of the wave action equation is used in all spectral wave models at global and regional scales, such as the WAve Model (WAM) by the WAve Model Development and Implementation (WAMDI) Group (WAMDI 1988); WAVEWATCH III by NOAA/NCEP, in the same spirit as the WAM model (Tolman 1991, Wavewatch III Development Group 2016); the Simulating WAves Nearshore (SWAN) model (Booij & Holthuijsen 1987, Holthuijsen 2007); or the TOMAWAC model of wave propagation in coastal areas developed by Eléctricité de France's Studies and Research Division (Benoit et al. 2001). Since some of these models, such as WAVEWATCH III and WAM, are being developed exclusively for wave modelling and thus are evolving independently from hydrodynamic and morphodynamic models, they will not be discussed further. Other wave models, such as SWAN and TOMAWAC, are integrated into the Delft3D and the TELEMAC modelling suites, respectively, for other applications. Another model that also used a spectral wave formulation is the XBeach model (Roelvink et al. 2009). The nonhydrostatic version of XBeach resolves the phases of the long waves and was first developed as a prototype version of the Simulating WAves till SHore (SWASH) model by Delft Technical University (Zijlema et al. 2011). SWASH has not yet been coupled with bedload transport or morphodynamics, and thus it will not be discussed further in this book either. See http://swash.sourceforge.net/features/features.htm for further details on SWASH.

3.2.1.1   XBeach

The model developed by Roelvink et al. (2009), commonly known as XBeach, consists of a nonlinear shallow-water model (NLSW) coupled with a model of propagation for the short-wave envelope and sediment transport and bed update formulations. It was originally developed for modelling the impact of storms with timescales of the order of days, on sandy coasts in model domains with dimensions on the order of kilometres (Roelvink et al. 2015), but has also been used for other applications. In XBeach, the model is set up so that the x-coordinate is the cross-shore direction and the y-coordinate is the alongshore direction, as shown in Figure 3.7 XBeach reproduces short- and long-wave processes. The short-wave processes that are included are refraction, shoaling, and breaking. Note that diffraction is a

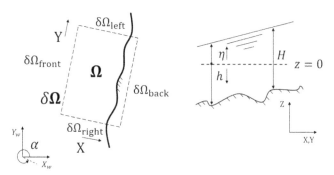

FIGURE 3.7
XBeach model set-up.

not well resolved by any spectral wave model formulation and thus not well resolved by XBeach; however in the nonhydrostatic mode of XBeach, this is improved; this formulation was initially developed by Delft Technical University (Zijlema et al. 2011). The long (infragravity) wave processes include generation, propagation, and dissipation of infragravity waves; wave-induced set-up and unsteady currents; and overwash and inundation. Sediment transport formulations for bedload and suspended load are included. The beach morphodynamic processes include dune face avalanching, bed update, and breaching.

The XBeach model can be set up as a one-dimensional (1D) cross-shore profile or a two-dimensional horizontal (2DH) model. In the 1D cross-shore model, one can retain some directional spreading or use one directional bins. In the 2DH mode, the model can be run in a rectilinear or a curvilinear staggered grid. In 2DH models, one can either propagate the waves in directional space and solve wave refraction 'on the fly' or find the wave direction at regular intervals with the stationary solver and then propagate the wave energy along the mean wave direction (Roelvink et al. 2015). The second option is better because it preserves wave groupiness and leads to more infragravity wave forcing.

The hydrodynamic module can be implemented in three different ways:

- the short wave variations and the long waves associated with them are resolved on the wave group scale—this is the original model set-up, described in Roelvink et al. (2009), corresponding to the surfbeat model;

- stationary wave model, which resolves wave-averaged equations with no infragravity waves; and

- nonhydrostatic (wave-resolving) model, which solves a combination of the NLSW equations with a pressure correction term.

The short waves are modelled with the wave action balance equation, and the wave–current interactions are into consideration through a correction on the wave number k using the Eikonal equations. These interactions are important for gully and rip-current applications (Reniers 2004). Wave dissipation in XBeach may be caused by wave breaking, bottom friction, and dissipation. Wave breaking is implemented through five different formulations, three for the standard hydrodynamics model and two for the stationary wave case. Dissipation due to bottom friction is particularly important in coral reef applications, and a spatially varying bottom friction coefficient needs to be implemented in such cases (Lowe et al. 2007, Van Dongeren et al. 2013, Quataert et al. 2015). Vegetation also affects short-wave dissipation. Following Mendez & Losada (2004), who considered vertically heterogeneous vegetation (for example, a dense root system but a sparse stem area), in XBeach, the dissipation due to the vegetation is assumed to depend on the local wave height and a number of vegetation parameters. These parameters are defined at different locations on the seabed. Wave breaking induces radiation stresses and turbulent fluctuations; these are both responsible for some of the sediment transport processes near the bed. The effect of turbulence near the bed can be computed in terms of the turbulence variance at the water surface, k, and the root-mean-square wave height, H_{rms}. Assuming an exponential decay model for the turbulence variance with a mixing length proportional to H_{rms}, the turbulence variance at the bed, k_b, may be defined as (Roelvink & Stive 1989)

$$k_b = \frac{k}{\exp\left(h/H_{\mathrm{rms}}\right) - 1}. \tag{3.6}$$

Finally, the roller energy from short waves is also computed, by coupling the roller energy balance to the wave-action balance equation, where dissipation of wave energy serves as a source term for the roller energy balance.

The low-frequency waves and mean flows are determined using a Lagrangian formulation of the shallow-water equations. In the generalized Lagrangian mean (GLM) framework, the velocities are expressed as the sum of Eulerian and Stokes-drift velocities (Andrews & Mcintyre 1978):

$$u^L = u^E + u^S, \quad \text{and} \quad v^L = v^E + v^S, \tag{3.7}$$

where the Stokes-drift components,

$$u^S = \frac{E_w \cos\theta}{\rho h c} \quad \text{and} \quad v^S = \frac{E_w \sin\theta}{\rho h c}, \tag{3.8}$$

depend on the wave-group short-wave energy, E_w, and direction, θ.

Additional details on the XBeach equations and parameter formulations can be found in the XBeach manual. Some formulations have been improved. From the sediment transport perspective, for example, the infiltration modelling approach (Packwood 1983) and a new morphodynamic evolution equation based on the sediment transport equation induced by waves and currents (Soulsby & Damgaard 2005) were added to XBeach, in order to improve the model's gravel and composite beach dynamic evolution predictions in complex environments (Jamal et al. 2012, 2014, McCall et al. 2014). The gravel beach models have been found to predict well the evolution of gravel beaches subject to overtopping events and the evolution of composite beaches in natural environments.

3.2.2 RANS Models

Reynolds-averaged Navier–Stokes (RANS) models are a form of the (momentum transport) Navier–Stokes equations. The Navier–Stokes equations, along with the mass conservation equation (commonly known as the continuity equation), are the governing equations for Newtonian fluids. In a Newtonian fluid, the viscous stresses are proportional to the rate of change of deformation. In an incompressible fluid, the density does not vary, and hence the instantaneous continuity equation can be shown to be equivalent to

$$\nabla \cdot \mathbf{u} = 0, \tag{3.9}$$

with $\mathbf{u} = (u, v, w)$ and where D/Dt is the material derivative. The momentum conservation equations, when pressure gradients and diffusivity are in equilibrium with the inertial terms, may be expressed as

$$\frac{\partial u}{\partial t} + \nabla \cdot (u\mathbf{u}) = -\frac{1}{\rho}\frac{\partial p}{\partial x} + \nu\nabla^2(u), \tag{3.10}$$

$$\frac{\partial v}{\partial t} + \nabla \cdot (v\mathbf{u}) = -\frac{1}{\rho}\frac{\partial p}{\partial x} + \nu\nabla^2(v), \tag{3.11}$$

$$\frac{\partial w}{\partial t} + \nabla \cdot (w\mathbf{u}) = -\frac{1}{\rho}\frac{\partial p}{\partial x} + \nu\nabla^2(w). \tag{3.12}$$

To illustrate how the RANS equations are obtained, we express Equations 3.9–3.12 in terms of the mean, $\bar{\mathbf{u}}$, and the fluctuating, $\mathbf{u}'$, velocity components; i.e., we do a *Reynolds decomposition* of the velocity and the pressure fields. Generally, for a flow variable ϕ, this means

$$\phi = \bar{\phi} + \phi'. \tag{3.13}$$

The fluctuating, turbulent motions are statistically random, and therefore they can be characterized with statistical concepts. The mean, on the other hand, is simply the integral over a time interval T or the average of discrete, equally spaced measurements:

$$\bar{\mathbf{u}} = \int_t^{t+T} \mathbf{u}\,dt = \frac{1}{N}\sum_{k=1}^{N}\mathbf{u}_k.$$

The turbulent fluctuations can then be defined very simply as the difference between the instantaneous velocity field and the velocity average:

$$\mathbf{u}' = \mathbf{u} - \bar{\mathbf{u}}.$$

Note that, by definition, the time averages of the velocity fluctuations are identically zero. In fact, a number of rules govern time-averaged flows. Say we have two flow variables, ϕ and ψ, with a mean component and a fluctuation component: $\phi = \bar{\phi} + \phi'$ and $\psi = \bar{\psi} + \psi'$. If $\bar{\bar{\phi}}$ is the fluctuation component of $\bar{\phi}$, then these two variables satisfy the following conditions: $\overline{\psi'} = \overline{\phi'} = 0$, $\bar{\bar{\phi}} = \bar{\phi}$, $\frac{\overline{\partial \phi}}{\partial s} = \frac{\partial \bar{\phi}}{\partial s}$, $\int \phi ds = \int \bar{\phi} ds$, $\overline{\phi + \psi} = \bar{\phi} + \bar{\psi}$, and $\overline{\psi \phi} = \bar{\phi}\bar{\psi} + \overline{\psi'\phi'}$.

By substituting the Reynolds decomposition into the Navier–Stokes equations (Equations (3.9) to (3.12)), taking the time average, and applying the rules for Reynolds decomposition, the RANS equations are obtained:

$$\nabla \cdot \bar{\mathbf{u}} = 0, \tag{3.14}$$

$$\frac{\partial \bar{u}}{\partial t} + \nabla \cdot (\overline{u\mathbf{u}}) = -\frac{1}{\rho}\frac{\partial \bar{p}}{\partial x} + \nu \nabla^2(\bar{u}) - \frac{1}{\rho}\left[\frac{\partial(\rho \overline{u'^2})}{\partial x} + \frac{\partial(\rho \overline{u'v'})}{\partial y} + \frac{\partial(\rho \overline{u'w'})}{\partial z}\right], \tag{3.15}$$

$$\frac{\partial \bar{v}}{\partial t} + \nabla \cdot (\overline{v\mathbf{u}}) = -\frac{1}{\rho}\frac{\partial \bar{p}}{\partial x} + \nu \nabla^2(\bar{v}) - \frac{1}{\rho}\left[\frac{\partial(\rho \overline{u'v'})}{\partial x} + \frac{\partial(\rho \overline{v'^2})}{\partial y} + \frac{\partial(\rho \overline{v'w'})}{\partial z}\right], \tag{3.16}$$

$$\frac{\partial \bar{w}}{\partial t} + \nabla \cdot (\overline{w\mathbf{u}}) = -\frac{1}{\rho}\frac{\partial \bar{p}}{\partial x} + \nu \nabla^2(\bar{w}) - \frac{1}{\rho}\left[\frac{\partial(\rho \overline{u'w'})}{\partial x} + \frac{\partial(-\rho \overline{v'w'})}{\partial y} + \frac{\partial(\rho \overline{w'^2})}{\partial z}\right]. \tag{3.17}$$

In the equations above, the mean inertia terms are not any more in balance with the pressure gradients and mean diffusion terms, as in the exact Navier–Stokes equations (3.9) to (3.12). Now, we have an additional term explicitly associated with nonlinear velocity fluctuations, which physically correspond to turbulent momentum fluxes; these terms are defined as *Reynolds stresses*, as a momentum flux is equivalent to a stress. The six unknown Reynolds stress components can be expressed in tensor form as

$$\tau_{ij} = -\rho \overline{u'_i u'_j}, \tag{3.18}$$

where τ_{ij}, the Reynolds stress tensor, is symmetric. With the flow and pressure fields now split into mean and fluctuating components, the conservation equations (3.14) to (3.16) have ten unknowns, and therefore, we need six additional equations to close the system and be able to determine both the mean and turbulent terms. This is called the *turbulence closure problem*: to close the system, some of unknowns need to be estimated using approximations of known flow properties. In some way, this means that Reynolds averaging causes a loss of the information contained in the Navier–Stokes equations regarding the exact properties of the flow (Brown 2016).

There are a number of turbulence closure approaches based on either of the following flow properties: the eddy viscosity, the mean kinetic energy, the turbulent kinetic energy, vortex stretching, or energy cascading. Some of them will be discussed below. To facilitate the analysis, the overbar of the mean quantities will be dropped, i.e. $\bar{\psi} = \psi$. The eddy viscosity turbulence models may assume that the normal Reynolds stresses are isotropic, which is usually inaccurate. These models are based on Newton's law of viscosity for incompressible (Newtonian) fluids, which states that the viscous stresses are proportional to the rate of deformation or the *strain tensor*, $\mathbf{S}$,

$$\tau_{ij} = \mu_t \mathbf{S} = \mu \frac{1}{2} \left(\frac{\partial u_i}{\partial x_j} + \frac{\partial u_j}{\partial x_i} \right), \qquad (3.19)$$

where $\mu_t = \nu_t/\rho$ is the dynamic eddy viscosity and ν_t is the kinematic eddy viscosity. However, in this approximation, we need to ensure that the normal Reynolds stresses are equal to the turbulent kinetic energy per unit volume, $\frac{1}{2}(\overline{u'^2} + \overline{v'^2} + \overline{w'^2})$, and therefore we need an extra term of the form $-2\rho k \delta_{ij}/3$,

$$\tau_{ij} = -\rho \overline{u'_i u'_j} = \mu_t \left(\frac{\partial u_i}{\partial x_j} + \frac{\partial u_j}{\partial x_i} \right) - \frac{2}{3} \rho k \delta_{ij}, \qquad (3.20)$$

where k is the turbulent kinetic energy.

For the mean kinetic energy turbulence closure scheme, the RANS momentum conservation Equations (3.15) to (3.16) are each multiplied by the mean flow components u, v, and w, respectively, and then added together and simplified to obtain an equation for the mean flow kinetic energy, K, of the form

$$\frac{\partial(\rho K)}{\partial t} + \nabla \cdot (\rho K \mathbf{u}) = \nabla \cdot (-p\mathbf{u} + 2\mu_t \mathbf{u} \mathbf{S} - \rho \overline{u'_i u'_j} \mathbf{u}) - 2\mu_t \mathbf{S} : \mathbf{S} + \rho \overline{u'_i u'_j} \cdot \mathbf{S}. \quad (3.21)$$

There are four divergence terms in the equation above, and they represent (from left to right) transport by advection of mean kinetic energy, pressure, viscous stresses, and Reynolds stresses. The first term on the left-hand side is the temporal gradient of the mean kinetic energy, while the last two terms on the right-hand side represent dissipation of mean kinetic energy due to 1) viscous stresses and 2) turbulent production, respectively.

For the turbulent kinetic energy turbulence closure scheme, the conservation equation for k is very similar to the one for the mean kinetic energy. The details on how to obtain the equation can be found in Tennekes & Lumley (1972), and it is of the form

$$\frac{\partial(\rho k)}{\partial t} + \nabla \cdot (\rho k \mathbf{u}) = \nabla \cdot (-\overline{p'\mathbf{u'}} + 2\mu_t \overline{\mathbf{u'} \mathbf{S'}} - \frac{\rho}{2} \overline{u'_i \cdot u'_i u'_j}) - 2\mu_t \overline{\mathbf{S'} : \mathbf{S'}} - \rho \overline{u'_i u'_j} \cdot \mathbf{S}.$$
$$(3.22)$$

This equation also contains transport terms associated with advection, pressure, viscous stresses, and Reynolds stresses, as well as dissipation due to viscosity and dissipation due to turbulence. The last term has a negative sign, in contrast with the positive sign that it has in Equation 3.21, because here it represents dissipation of turbulent kinetic energy.

The last two turbulence closure schemes considered are the vortex stretching and the energy cascade schemes. Both methods are fundamental to how energy is passed between different turbulence scales. Vortex stretching produces a change in vorticity in the direction of the stretching, in order to conserve angular momentum. Vortex stretching also produces energy cascading from large to small eddies, because the small eddies are exposed to the rate of strain of the large eddies, and therefore a change in vorticity, and a flux of energy from large eddies to small eddies. Most of the energy taken by an eddy of a given size is taken from the next largest eddy in size and passed to the next smallest eddy in size. This is equivalent to a cascading waterfall, where a filling pool overflows into the next pool, and this explains the energy cascade term so commonly used in turbulence dynamics. As the eddies become smaller and smaller, viscous stresses become more and more important until they completely dominate the dynamics and the eddy energy is dissipated as heat. The smallest scale known is the *Kolmogorov microscale* and is denoted here as η_k. Thus, the turbulent energy spectrum, $E(k)$, can be divided into three regions, each dependent on the size of the wavenumber. The range of small wavenumbers is the energy production subrange, which contains most of the energy and is dominated by large eddies. The range of intermediate wavenumbers is the dissipation subrange, and in this range, energy cascading occurs from large to small eddies. This is the inertial subrange, and the energy decays with increasing wavenumber at a rate that is inversely proportional to $k^{5/3}$, where k is the wavenumber, and proportional to $\epsilon^{2/3}$, where ϵ is the energy dissipation rate:

$$E(k) = \alpha \epsilon^{2/3} k^{-5/3}. \tag{3.23}$$

The dissipation subrange corresponds to the Kolmogorov microscale; here all the mechanical energy remaining in the system is converted into thermal energy through frictional forces. In nonlinear systems, an inverse energy cascade may take place. When all turbulent scales are resolved, the numerical simulation is referred to as a direct numerical simulation (DNS). Contrary to DNS, in RANS, the subgrid-scale turbulence is not resolved but is approximated using either one of the turbulence closure schemes described above. A Reynolds-averaged simulation (RAS) is a simulation where the RANS equations are solved. A large-eddy simulation (LES) is a compromise between an RAS and a DNS. In LES, the equations are filtered so they can solve the large eddies in the flow. As LES may present some issues near boundaries, in some cases, hybrid methods are developed using DNS near walls and LES far from them.

A number of computational fluid dynamics (CFD) models are available to solve RANS, LES, and DNS problems. Some of them are commercial and some are open source software. Examples of proprietary software include the products developed by ANSYS, such as Fluent or CFX (see www.ansys.com/products/fluids for further details). ANSYS Fluent can be used for turbulence modelling and multiphase flows, as well as fluid–structure interactions, and thus can be useful for modelling in coastal environments.

CFX is designed for mechanical engineering applications, for example, for turbomachinery or for gas or hydraulic turbines, and is not really appropriate for morphodynamics modelling. A third example of proprietary software is STAR−CCM+, which has been used as well for fluid–structure interaction analyses. An example of a CFD model in extensive use is the Open source Field Operation And Manipulation (OpenFOAM®) CFD toolbox, for which a number of sediment transport modules have been developed. However, only the hydrodynamic models are widely available (see www.openfoam.com/), and this applies to most open source CFD codes, with authors having to design most of their sediment transport and morphodynamic routines in-house. The biggest problem with CFD models is that they are computationally expensive, so they are only suitable for small timescales and space scales.

Mesh-based computational techniques with explicit time marching algorithms need to satisfy the Courant–Friedrichs–Lewy (CFL) stability criterion. The CFL criterion is expressed in terms of the *Courant number*, C_o. For a space scale Δx, a timescale Δt, and a characteristic speed u, C_o should be equal or less than $C_{o,\max}$, i.e.,

$$C_o = \frac{u\Delta t}{\Delta x} \leq C_{o,\max}. \tag{3.24}$$

To capture all the turbulent scales, one needs to use DNSs (Vittori & Verzico 1998), which is in most cases impractical. Indeed, for regional modelling, DNS models tend to be too computationally expensive and therefore will not be discussed any further. However, if such model is needed, some open source options are available (Popinet 2003). RANS models, on the other hand, are more computationally efficient and capture well the mean and turbulent flows in the surf zone through turbulent closure schemes, both in mild and in energetic conditions (Brown et al. 2016).

RANS models have been developed to analyse a number of sediment transport and morphodynamic problems. We first discuss the application of OpenFOAM to the study of hydrodynamics and sediment dynamics over a nearshore sandbar (Jacobsen et al. 2014). First, sediment transport formulations are assessed as a function of the surf similarity parameter, ξ_o (Battjes 1974),

$$\xi_o = \frac{\tan(\beta)}{\sqrt{H_o/L_o}}, \tag{3.25}$$

and two expressions for Dean's parameter, the regular one, denoted as Ω_D, and defined as (Dean 1973)

$$\Omega_D = \frac{H_B}{w_s T}, \tag{3.26}$$

and the slope corrected Dean's parameter, Ω_{HK}, defined by Hattori & Kawamata (1980) as

$$\Omega_{HK} = \Omega_D \tan(\beta). \tag{3.27}$$

The surf similarity parameter ranges between 0.08 and 1.19, Ω_D between 1.1 and 27.7 and Ω_{HK} between 0.01 and 0.52. H is the wave height, T the wave period, and w_s the settling velocity based on the median grain diameter, d_{50}; $\tan(\beta)$ is the slope of the beach. As is customary, H_o denotes wave height in deep water, and H_B wave height at breaking depth. In the first part of the study, the bed is fixed, while in the second part (Jacobsen & Fredsøe 2014), the bed is allowed to evolve. As mentioned above, the model is not open source, but in the first part of the paper, the authors describe in detail the approach they adopted so that mass conservation is achieved.

Mass conservation for bedload and suspended load is a critical issue, which has to be addressed in all sediment transport models. For example, a discrete-vortex, particle-tracking model developed during a project funded by the European Commission between 2002 and 2005 (contract no. EVK3-2001-00056), on Sand Transport and Morphology of Offshore Sand Mining Pits/Areas (the SANDPIT project), to analyse sediment dynamics above rippled beds, although validated against experimental data (Van der Werf et al. 2008), required extensive adjustments before the results could be confidently presented in a further work (Malarkey et al. 2015). In the research by Jacobsen and collaborators, they follow a similar development path, and once the mass conservation procedure has been validated, they develop the model further for morphodynamic studies. Using a wide range of the surf similarity parameter (Equation 3.25) allows them to consider spilling and plunging breaker cases. Dean's parameter (Equation 3.27), on the other hand, is used to classify different breaker bar types. The cross-shore evolution of breaker bars is driven by different cross-shore coastal processes, including nonlinear wave effects, such as Stokes drift, wave asymmetry, and skewness, and wave-driven currents such as streaming and undertow. According to Jacobsen et al. (2014), the undertow is a crucial process in suspended sediment dynamics. Undertow is caused by the depth-varying shear stress, combined with an overall balance between a net shoreward volume flux above the wave crest and a seaward volume flux below the wave trough. In a more recent work, Roelvink & Stive (1989) showed that the wave asymmetry drives the breaker bar shoreward, whereas the undertow drives it seaward.

Because RANS models using the Volume-of-Fluid method to describe the free surface are wave-phase-resolving methods, and because most suspended sediment methods developed for RANS models are sediment-phase-resolving as well, it is possible to analyse the effects of wave asymmetry, wave skewness, wave breaking, or phase lag between the maximum wave height and maximum bed shear stresses, amongst others (Jacobsen et al. 2014). Contrary to the discrete-vortex, particle-tracking model (Malarkey et al. 2015) where the bed is assumed to have an equilibrium fixed shape, Jacobsen et al. (2014) coupled the hydrodynamic and sediment transport equations with a bed-level updating scheme and were able to analyse the evolution of the bed and correct the width and height of the breaker bar at every timestep of the simulation. However, during the numerical model development process, it is very important that each of the capabilities of numerical models is tested, through benchmark tests that increase in difficulty incrementally. Jacobsen et al. (2014) test first the model with a fixed, flat bed with constant slope, then with a fixed barred profile. In their second paper, Jacobsen & Fredsøe (2014) discussed the sediment transport and morphodynamic implementation. The bed-level change is a solution of the Exner equation, with contributions to the bulk cross-shore sediment transport rates from both bedload and suspended load. The sediment transport formulation of Jacobsen & Fredsøe (2014) is phase resolved, so that the onshore and the offshore contributions to sediment transport can be split and analysed. This is performed with a fixed bed, in order to understand the sediment dynamics predicted by model. They also consider the morphological evolution of a cross-shore profile with an initial constant slope and test the breakpoint hypothesis of sandbar evolution. This breakpoint hypothesis, in short, states that shoreward migration occurs in long periods of mild, non-breaking wave conditions, while seaward migration occurs in short periods of high-wave-energy conditions. From a series of experiments with different grain sizes and fixed or mobile beds, Jacobsen & Fredsøe (2014) are able to demonstrate that the surf similarity parameter and the modified Dean's parameter determine the cross-shore suspended sediment transport rate. The surf similarity parameter is shown to control the location of the maximum in the seaward transport rate, and the modified Dean's parameter controls the magnitude of the maximum seaward transport rate. In relation to the breaker bars, these are shown to reach a quasiequilibrium profile, where the crest height adjusts to the local water depth. Breaker bar migration is always offshore under the action of the specific wave that led to its formation. The morphological response is smoothed as the incident wave deviates from a regular wave. Finally, the morphological response is stronger with steeper initial slopes.

OpenFOAM has different turbulence-closure schemes already implemented in the open source version of the code, which have been assessed extensively by Brown (2016). In this research, Brown (2016) assessed four different types of turbulent-closure schemes. They consist of two equations in which an eddy viscosity is used to calculate the Reynolds stress. The first equation is for the turbulent kinetic energy (TKE), k. The second equation is either for the turbulence dissipation rate, ϵ (k-ϵ models), or a characteristic frequency, ω, associated with the turbulence (k-ω models). The turbulence-closure models that were assessed were the k-ω model described by Wilcox (1993), the k-ω shear stress transport (SST) model developed by Menter (1994), the renormalised group (RNG) k-ϵ model (Bradford 2000), the nonlinear k-ϵ model developed by Shih et al. (1996), and a full Reynolds stress model (RSM) developed by Launder et al. (1975). For this analysis, Brown et al. (2014) developed a new library of turbulence-closure models for *OpenFOAM* which included variations of the density in the free surface, an idea inspired by the work of Jacobsen (2011). When comparing the model predictions with experimental data, the model that gave the best Brier skill score was the nonlinear $k - \epsilon$ model of Shih et al. (1996), followed in second place by the full RSM of Launder et al. (1975), then in third place by the k-ω SST model, in fourth place by the k-ω model, and in fifth (and last) place by the RNG k-ϵ model. The nonlinear k-ϵ model was able to reproduce as well a number of turbulence characteristics associated with different types of breakers.

The RANS model in OpenFOAM is solved using a finite volume discretization scheme, with a collocated variable arrangement. Spatial discretization schemes for CFD and regional shallow-water models include finite difference, finite element, finite volume, and spectral or boundary element schemes. Conceptually, in finite difference schemes, the derivatives in the partial differential equations are approximated by linear combinations of function values at the grid points; this is the oldest discretisation scheme. Finite difference schemes can be implemented in both structured and unstructured grids, but generally finite difference methods are implemented in structured grids, which are topologically equivalent to Cartesian grids. The finite volume scheme is the most common discretisation scheme in transport problems, for a number of reasons. Firstly it is readily applied to unstructured grids that fit complex geometries. The finite volume method (FVM) leads to a finite-difference-like discretisation on a Cartesian grid for the underlying, integral surface and volume conservation laws. In a finite element method, one has to choose the shape functions (elements) and weighting functions. Spectral or boundary element methods will not be discussed here.

Some OpenFOAM module developments for sediment transport and morphodynamics have been undertaken by a number of researchers, but the pace of development has been much slower than that for the hydrodynamic counterparts. This is because the hydrodynamics need to be well understood and validated in the models before the sediment and morphodynamic modules are developed. There are fewer researchers working on the latter than on the former, and those working on both usually publish the hydrodynamics research a couple of years before the sediment transport or the morphodynamics results and codes are released. Here we will describe the morphodynamics research by Jacobsen & Fredsøe (2014) and a sediment transport module for OpenFOAM developed by Brown et al. (2020, under review).

Jacobsen & Fredsøe (2014) solve an advection–diffusion equation with a settling sediment term, as described by Fredsøe & Deigaard (1992) and also Engelund & Fredsøe (1976), and a Schmidt number of 1, which means the sediment diffusivity is equal to the eddy viscosity. This is not necessarily always true, in particular in the presence of bed ripples, as discussed by Davies & Thorne (2008) and by Malarkey et al. (2015). However, Jacobsen & Fredsøe (2014) did not take this effect into account. As they were studying the evolution of breaker bars, and such evolution occurs at Shields parameters larger than 1—equivalent to a nondimensional transport rate above 12.5 under oscillatory flow (Madsen & Grant 1977)—there is no need to include corrections for unresolved bed features in the sediment concentration conservation equation, nor in the Exner equation for bed-level evolution, because the model design will limit the growth of small bed features.

The Exner equation that Jacobsen & Fredsøe (2014) used has bedload and suspended load contributions. Two cases were considered, one involving sediment transport over fixed beds and one case which considers seabed evolution. For the fixed bed case, the surf similarity parameter ξ_o, Dean's parameter Ω_D, and the slope-corrected Dean's parameter Ω_{HK}, defined above, are used to describe maximum transport rates and bulk transport patterns. Ω_D is a beach shape (or beach state) classification parameter (Wright & Short 1984), while Ω_{HK} is an indicator of bulk beach erosion and accretion (Hattori & Kawamata 1980). The RANS model is solved with the $k - \omega$ turbulence-closure scheme, with k the turbulent kinetic energy, and ω a characteristic frequency of the flow. Both quantities are defined in terms of the rotation tensor of the velocity field. The resulting equations are solved using the finite volume discretisation scheme, based on the integral form of the equations, with collocated variables. The suspended sediment concentration conservation equation that is solved has the form

$$\frac{\partial C}{\partial t} + \nabla \cdot \{[\alpha \mathbf{u} + \mathbf{w}_s(\mathbf{x})] C\} = \nabla \cdot [\alpha (\nu + \nu_t) \nabla C], \qquad (3.28)$$

where $\mathbf{w}_s(\mathbf{x}) = w_s(0,0,1)$ is the settling velocity vector, C the volumetric suspended sediment concentration. The volume of fluid (VOF) approach captures the free surface (Berberović et al. 2009) through the variable α, which takes a value of 1 in water and a value of 0 in air. A value of α between 0 and 1 indicated the volume of fluid in the numerical fluid parcel, taking into consideration that the two phases (air–water) are immiscible. ν and ν_t are the molecular and turbulent viscosities, respectively. The inclusion of ν ensures numerical stability. Also, since α is not multiplying the temporal gradients of C in Equation (3.28), the sediment is kept within the water phase of the two-phase flow (Jacobsen et al. 2014).

Suspended sediment transport is uncoupled from the bed-level change, but bedload usually is coupled with the bed updating algorithm and is dominant at regional scales, bedform evolution studies—as in the case of sandbank evolution discussed in Chapter 2. In the cross-shore study of Jacobsen et al. (2014), the period-averaged, depth-integrated suspended sediment flux may be expressed as

$$\overline{q_s} = \int_{-h}^{\eta} \overline{uC + \widetilde{u}\widetilde{C}} dy, \qquad (3.29)$$

i.e., as the sum of the period-averaged sediment advection term and the oscillating sediment advection term. Note that $\widetilde{C} = C - \overline{C}$, and therefore it is not necessary to split the suspended sediment transport flux into two components. Here u is the horizontal velocity, h is the local water depth, and η is the local water elevation.

In the breaker bar evolution case analysed by Jacobsen et al. (2014), which was previously discussed, the bedload transport plays a role in the Exner equation for the bed evolution, which in period-averaged form is expressed as

$$\frac{\partial \overline{z_b}}{\partial t} = \frac{1}{1 - e_d} \left[\nabla \cdot \overline{q_b} + \overline{E} + \overline{D} \right], \tag{3.30}$$

where e_d is the sediment porosity, q_b the bedload transport, and E and D the erosion and deposition parameters associated with the suspended sediment transport. As before, overline means time averaging over one wave period, and z_b is the vertical coordinate of the seabed.

3.2.3   Boussinesq-Type Models

Boussinesq-type models (BTMs) are models of another type for the analysis of long ocean waves in shallow waters. These models are suitable for engineering design and environmental management activities (Brocchini 2013), such as design of offshore and nearshore structures, pipelines, underwater cables, or coastal defences; morphodynamics modelling and pollutant dispersion; and riverine engineering. The seminal mathematical work on modelling long waves in shallow water by Boussinesq dates back to 1877 (Boussinesq 1877), where he proposed a theory explaining the observations of solitary waves by J. C. Russell. The equations proposed by Boussinesq are known as the Korteweg–de Vries (KdV) equations, after the two researchers who rediscovered them at the end of the 19th century (Darrigol 2005). Since Boussinesq, a number of researchers in coastal engineering and the geosciences, such as Howard Peregrine from the University of Bristol (Peregrine 1972); Ib Svendsen from the University of Delaware (Svendsen 1984); Per Madsen from DTU (Madsen & Sørensen 1992); James Kirby from the University of Delaware, who developed FUNWAVE based on the model by Wei et al. (1995); and Maurizio Brocchini from Ancona University, who did his PhD with Howard Peregrine and went on to become a distinguished researcher in this field. The regional model MIKE21, from the Danish Hydraulics Institute (DHI), is based on the Boussinesq-type equations (BTEs) but with additional elements which complete a BTM. The elements that Brocchini (2013) cites as part of BTMs are the following:

- a proper account of the physics, including the propagation of the waves and fluid–sediment interactions, amongst others;

- a choice of the most suitable model equations, either with a good description of natural environments or idealized model set-ups;

- an adequate definition of the model boundary conditions;

- the use of the most suitable numerical approach, for the chosen model equations and boundary conditions.

There are two classes of BTEs. The first class (1) assumes that the flow is irrotational, while the second class (2) takes into account rotationality and turbulence within the flow. Class (1) models are appropriate for deep to intermediate flows, and dissipative mechanisms associated with wave-breaking and bottom friction have to be added *a posteriori*, and somewhat artificially, which is far from ideal, even though the 'surface roller' concept for surface wave dissipation has been implemented quite successfully in a number of models. Examples of class (1) models include the DHI model MIKE21 (Madsen & Sørensen 1992) and Delaware's FUNWAVE model (Wei et al. 1995). Class (2) models include dissipative mechanisms induced by wave breaking, and these models seem to be better for surf zone flows (Veeramony & Svendsen 2000). Because of the derivation procedure, model equations require extra closures for both vorticity and turbulence, which can be calculated either with semianalytical or numerical methods. Class (2) models have a scientific interest but are less popular amongst practitioners, due to the difficulties associated with the vorticity and turbulence-closure schemes and the numerical implementation of such models. The advantage of BTMs over NLSW ones is that propagation of waves near obstacles and harbours are better reproduced, because BTMs can represent better the wave diffraction processes.

3.3 Modelling Complex Environments

3.3.1 Some Generalities

While many environments may be wave dominated or tide dominated, the large majority evolve, thanks to the combined action of both waves and tides. The relative importance of these forcings depends, partially, on the range of values of the Irribaren number and the Dean parameter, which include hydrodynamic, bed slope, and grain size controlling mechanisms. Other considerations, such as substrate and vegetation characteristics, for example, are also important. Numerical models can, in principle, predict the physiographic evolution of complex environments to a good extent, based on wave conditions and tidal regimes, such as seasonal changes of beaches from erosive (generally in winter) to accretive (generally in the summer) or dissipative (during spring tides) beach configurations, and this may be reproduced well with numerical models. If the models are designed with the correct resolution, in the order of metres for example, they can also reproduce more detailed morphodynamic evolution, such as shoreward propagating accretionary waves (SPAWs); these are coherent bed features that have been observed during obliquely incident wave conditions; they were first reported by Wijnberg & Holman (2007), who identified them from video observations.

To analyse coastal evolution in complex environments, we rely on regional models coupling wave, tide, and sediment transport modules and model parameters that incorporate the effects of spatially varying bottom roughness and vegetation. These coupled models work as 3D–2D–1D oceanographic modelling tools and are used to analyse ocean circulation, coastal current dynamics, wave propagation and transformation processes, wave–current interactions and feedback mechanisms, as well as the interactions of waves and currents with bedload, suspended load, and bed morphodynamics. Other applications include, for example, water quality monitoring and particle tracking, with an embedded biogeochemical module, to perform coastal dispersion studies. Some model developers include the model coupling tools, permitting interactions of the forcings.

For some applications, however, the module coupling is performed offline, for example, for water quality modelling. This assumes that the particles in suspension have little effect on the hydrodynamics, which, for chemicals and small noncohesive contaminants, is a good approximation. However, this does not hold in flows with high suspended sediment concentrations, where flow–sediment interactions have to be included. In estuarine environments, where temperature and salinity gradients could be important, it is possible to analyse thermohaline circulation and assess the behaviour of baroclinic, mixing flows, by including boundary conditions for temperature and salinity.

Generally, the conservation equations for constituents are advection–diffusion equations, for salinity, temperature, and biochemical constituents, with first-order decay for oil spills, or other degrading constituents. Some models, such as Delft3D, have a formulation for cohesive and noncohesive sediments, making them ideal options for both open beach and estuarine environments. Bed evolution is modelled with the Exner equation, which includes bedload transport processes through computation of the sediment transport rate. The seabed evolves at a slower rate than the hydrodynamic forcings. This means that the bed does not need to be updated at every hydrodynamic timestep. A morphodynamic updating factor, or MORFAC parameter, defines the number of skipped hydrodynamic timesteps between morphological updates. The model predictions need to be tested with different MORFAC parameter values to make sure that the results are still reliable with MORFAC larger than 1.

Regional models can have several scales. The scale determines whether the Coriolis forcing needs to vary with latitude or whether it may be assumed as constant. For local-scale models, a Cartesian coordinate system is adequate. However, for regional or global-scale models, the spherical coordinate system is a better choice, because it takes into account the change with latitude of the Coriolis forcing. Designing a mesh with the optimal characteristics is very important, not only to maximize the numerical stability of the models, but also to improve the precision of the model predictions. If the problem focuses on sediment transport and morphodynamics, then the terrain-following vertical coordinate system is the preferred choice. If, in contrast, the problem focuses on thermohaline circulation, then a z level coordinate system is preferred. This is because in z coordinates, the vertical layers have the same thickness, and errors associated with integration over layers with varying thickness are removed.

In macrotidal environments, an important consideration is to start the model during high tides. Although most models include now robust wetting and drying algorithms, the model is more stable when the first few timesteps take place in the drying part of the tidal cycle. Along the same lines, spin-up times need to be analysed and adjusted so that the computations of the fundamental and secondary variables are performed once the model has reached a stable solution. Some authors consider a full month for model spin-up, but if the model is based on a hydrodynamic module only, the model converges within a couple of days, at most. Longer spin-up times are possibly needed in baroclinic models, due to the thermohaline stratification. If the model is run from a hot start, that is, with a restart file from a previous computation, then a spin-up time is not necessary.

Finally, with the increases in temperature, sea level, and the frequency and intensity of extreme events, our tools need to be tested and improved under extreme-event forcing conditions. During extreme events, water levels and speeds are very high compared with the case of climatologically normal conditions, which implies that smaller cell sizes and smaller timesteps are required to satisfy the CFL stability condition. Although it is possible to use desktops for model simulations of only a few days duration, it is necessary to optimise the numerical models and design model versions that can run smoothly on high-performance computers. An example of an optimised model is Firedrake (Jacobs & Piggott 2015), developed by Imperial College (see www.firedrakeproject.org). However, it is still not widely used by industry, which can be a problem as practitioners are more likely to trust models which have been extensively tried and tested in practical situations. Other models, which include sediment transport and morphodynamic modules and have been tested by industry under hurricane conditions, such as XBeach or Delft3D for example, for the time being benefit from their wide acceptance within large practitioner communities and already include most of the physics required for the simulation of morphodynamics and sediment transport in the coastal zone (Roelvink & Walstra 2004). Modelling with data assimilation can also improve the accuracy and reliability of model predictions (Van Dongeren et al. 2008); however, this is not yet a widely adopted approach in coastal analysis.

3.3.2   The Mouth of the Columbia River: A Case Study

In order to illustrate some modelling procedures in complex environments, an example for the Mouth of the Columbia River (MCR) will now be presented (Elias et al. 2012). The MCR is a very energetic inlet with complex dynamics, with rapid rates of water flow, sediment transport, and morphological change. This is a challenging field site location for experimentation. However, a dataset comprising measurements of currents, waves, salinity, temperature, and sediment transport at different observation stations along a transect of the MCR was obtained during the summer of 2005, as part of the 'Mega Transect Experiment' (MTE) led by the United States Army Corps of

Engineers (USACE) in collaboration with the United States Geological Survey (USGS). This dataset, together with tidal gauge measurements along the river and data from an offshore wave buoy, is detailed enough to be used for calibration and validation of Delft3D, or any other process-based regional model. The Columbia river has been the subject of numerous studies for a number of reasons, including the following:

- it is the largest US river on the Pacific coast;

- its shipping lanes are regularly dredged;

- it forms vital river connections for salmon migration;

- it provides the connection for sediment exchange between the river basin and the Columbian River littoral cell; and

- it creates a buoyant plume of freshwater outflow into the Pacific.

An important contribution of the work by Elias et al. (2012) was to include in the process-based model the effect of waves and the 3D circulation at high resolution, in order to model the sand transport processes at the MCR with sufficient accuracy. This is particularly important when analysing the erosion processes of the adjacent coastline and the foundations of the inlet jetties defending the entrances of the Columbia River and Grays Harbor, shown in Figure 3.8. The jetties were built at the end of the 19th century, as a line of defence of the river and of the Harbor against incoming waves and storms (Gelfenbaum & Kaminsky 2010). The 165-km long Columbia River littoral cell (CRLC) and its subcells are shown in Figure 3.8, together with other coastal landmarks.

The dataset used for calibration and validation is discussed, before we describe the model set-up. Most of the instruments are shown in Figure 3.9, with the locations of the five tripods deployed during the 'mega-transect experiment' in the summer of 2005, labelled as MGT 1–5, being shown in the figure inset. The MGT experiment was designed to evaluate the sediment fluxes across the Columbia River mouth. The MGT data was collected between 3 August and 9 September 2005 by four of the five tripods. (MGT3 was moved from its deployment position by a barge and stopped working properly on 16 August 2005.) The tripods were equipped with ADCPs, Acoustic Doppler Profilers (ADPs), and Acoustic Doppler Velocimetres (ADVs) to measure waves and currents throughout the water column and near the seabed, at the appropriate resolution. The ADPs at MGT1 and MGT2 measured water velocities at 1 Hz every 30 min and at 1,500 and 500 kHz, respectively, during 5 min bursts, every over 30 min. ADCPs at MGT4 and MTG5 also measured water velocities at 1 Hz every 30 min. At all stations, 5 min ensembles were computed from the raw data. Waves were measured half hourly with 17 min bursts at 2 Hz, using a second profile setting configuration. The ADVs, on the other hand, were placed close to the bed, to measure near-bed currents and orbital wave velocities.

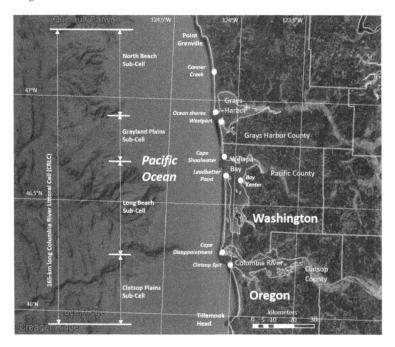

FIGURE 3.8
The Columbia River littoral cell and subcells, with geographical benchmarks.
(Modified from Gelfenbaum & Kaminsky 2010.)

Optical backscatter sensors (OBSs) and Microcat SBE CTD measurements were also performed to determine changes in suspended sediment concentrations, conductivity, temperature, and salinity. The OBSs were calibrated with sediment samples from the measurement sites in order to transform counts into kg/m^3. The weather and hydrological conditions encountered during the experiment were typical summer conditions: a relatively small freshwater daily inflow from the river with a discharge of 4000 m^3/s on average; mild winds with speeds varying between 0.2 and 9.2 m/s at 10 metres above ground level; and mild to medium wind wave conditions, with wave heights between 0.64 and 2.41 m, coming predominantly from the northwest. From the observations, the following conclusions were drawn:

- tidal ranges varied between 2 and 4 m, placing the Columbia River unambiguously within the mesotidal regime;

- tidal velocities reached speeds of around 2.5 m/s during spring tides, and spring–neap tidal velocity modulations coincided well with the spring–neap modulations of the water level;

- there were pronounced spring–neap variations in sediment transport rates and salinity;

- during strong opposing ebb currents, waves nearly doubled in height;

- the largest suspended sediment concentrations were observed during maximal spring tide velocities; and

- sediment transport was largest in the middle of the channel, where current velocities were also the largest.

- there was no significant wave breaking over the shoals, due to the mild summer wave conditions and the sheltering of the wave climate by the jetties;

- salinity variations at the MCR were very large, from 25 ppt during spring ebb tide to 33 ppt during mean tidal conditions;

- significant daily and semidiurnal variations in water temperature were observed; the variation covered a 7.5°C–15.5°C temperature range and was associated with the flood-ebb cycle.

Salinity and water-level stations—shown in Figure 3.9—along the river, obtained from the NOAA and the SATURN observation network, which can be consulted at http://www.stccmop.org/datamart/observation_network, were used for model calibration. The SATURN observation network is operated by the Center for Coastal Margin Observation and Prediction (CMOP), hosted by the Oregon Health and Science University (check www.stccmop.org/ for more information). Additional wave, wind, and river discharge observations were used as boundary conditions for the model: continuous offshore wave and wind data from the NDBC buoy no. 46029, located about 37 km west of the Columbia River entrance, and river discharge and water-level data from a USGS station upstream of the river at Beaver Army Terminal near Quincy, Oregon (not within river map boundaries of Figure 3.9).

The model developed by Elias et al. (2012) consisted of a 3D model that was capable of reproducing the dynamics and the sediment transport processes at the MCR under the influence of three forcings: the freshwater buoyant plume discharge from the Columbia River, the strong tidal currents moving in and out of the MCR, and the incoming wave climate from the Pacific Ocean. This was the first time the Delft3D model was tested under the combined influence of these three forcings at a highly energetic estuary mouth. The Delft3D-SWAN hydrodynamic, wave and sediment transport model set-up for sand-sized sediments was validated and calibrated using the five MGT observation points and the tidal gauges along the River. An important objective of the study was to have a realistic representation of the 3D currents in the waves module, because several authors have shown that 3D currents can have a significant effect on the wave field predicted by SWAN (Booij et al. 1999), particularly around tidal inlets (Van der Westhuysen et al. 2012). So the model consists of hydrodynamic, wave action, and constituent transport equations. The formulation was similar to other shallow-water regional models, with particular model configurations and model calibrations associated with the MCR

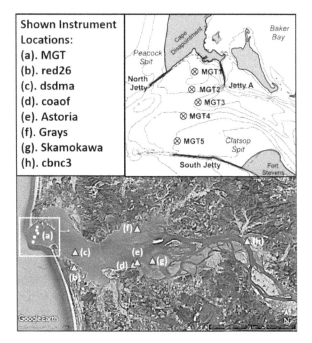

FIGURE 3.9
The Columbia River with some instrument locations at the mouth and along the river. (Modified from Elias et al. 2012.)

case study. Delft3D solves NLSW equations, which are stated here in 2D for simplicity, in the (x, y, σ) coordinate system. σ is the terrain-following vertical coordinate, and in Cartesian systems, it is defined as

$$\sigma = \frac{z - \zeta}{H},$$

where H is the total water depth (m) and ζ is the water elevation with respect to mean water level (MWL).

In shallow-water models, vertical accelerations are neglected, so the vertical momentum conservation equation reduces to the hydrostatic (or nonhydrostatic, depending on the formulation) pressure gradient balance equation, which is of the form

$$\frac{\partial P}{\partial \sigma} = -\rho g H, \tag{3.31}$$

where P is the (scalar) pressure field, ρ the water density (kg/m^3), and g the gravitational acceleration (m/s^2). The horizontal momentum conservation equations for U and V, on the other hand, may be expressed as

$$\frac{\partial U}{\partial t} + U\frac{\partial U}{\partial x} + V\frac{\partial U}{\partial y} + \frac{\omega}{h}\frac{\partial U}{\partial \sigma} - fV$$

$$= -\left[g\frac{\partial \zeta}{\partial x} + g\frac{h}{\rho_0}\int_\sigma^0\left(\frac{\partial \rho}{\partial x} + \frac{\partial \sigma'}{\partial x}\frac{\partial \rho}{\partial \sigma'}\right)d\sigma'\right]$$

$$+ \nu_H\left(\frac{\partial^2 U}{\partial x^2} + \frac{\partial^2 U}{\partial y^2}\right) + \frac{1}{h^2}\frac{\partial}{\partial \sigma}\left(\nu_V\frac{\partial U}{\partial \sigma}\right) + M_x \quad \text{and} \quad (3.32)$$

$$\frac{\partial V}{\partial t} + U\frac{\partial V}{\partial x} + V\frac{\partial V}{\partial y} + \frac{\omega}{h}\frac{\partial V}{\partial \sigma} + fU$$

$$= -\left[g\frac{\partial \zeta}{\partial y} + g\frac{h}{\rho_0}\int_\sigma^0\left(\frac{\partial \rho}{\partial y} + \frac{\partial \sigma'}{\partial y}\frac{\partial \rho}{\partial \sigma'}\right)d\sigma'\right]$$

$$+ \nu_H\left(\frac{\partial^2 V}{\partial x^2} + \frac{\partial^2 V}{\partial y^2}\right) + \frac{1}{h^2}\frac{\partial}{\partial \sigma}\left(\nu_V\frac{\partial V}{\partial \sigma}\right) + M_y, \quad (3.33)$$

respectively. ω is the vertical velocity component in the σ coordinate system, h is the water depth (m) from MSL, and f is the Coriolis coefficient (s^{-1}). U and V (m/s) represent the generalised Lagrangian mean (GLM) velocities: $U = u + u_s$ and $V = v + v_s$, with (u_s, v_s) being the Stokes drift (as in Equation 3.7), for simulations including waves (Lesser et al. 2004). (M_x, M_y) represent the wave and wind stresses on the surface of the sea. ρ_0 is a reference water density (kg/m^3). ν_H and ν_V (m^2/s) are the horizontal and vertical kinematic viscosities, respectively. The $k - \epsilon$ turbulence-closure model is used, where k is the turbulent kinetic energy and ϵ is the turbulent energy dissipation rate. In this framework, the vertical kinematic viscosity is expressed as

$$\nu_V = c'_\mu c_D \frac{k^2}{\epsilon},$$

where c'_μ and c_D are calibration constants.

The continuity equation is expressed in terms of the depth-averaged velocity components, $(\overline{U}, \overline{V})$, as a conservation equation for the water elevation ζ:

$$\frac{\partial \zeta}{\partial t} + \frac{\partial(h\overline{U})}{\partial x} + \frac{\partial(h\overline{V})}{\partial y} = S, \quad (3.34)$$

where S corresponds to any sources or sinks of mass per unit area.

Finally, the conservation equation for constituent c, either temperature (°C), salinity (kg/m^3), or sediment concentration (kg/m^3), depends on variables that have already been defined from previous equations and is as follows:

$$\frac{\partial(hc)}{\partial t} + \frac{\partial(hUc)}{\partial x} + \frac{\partial(hVc)}{\partial y} + \frac{\partial(\omega c)}{\partial \sigma} = h\left[\frac{\partial}{\partial x}\left(D_H\frac{\partial c}{\partial x}\right) + \frac{\partial}{\partial y}\left(D_H\frac{\partial c}{\partial y}\right)\right]$$

$$+ \frac{1}{h}\frac{\partial}{\partial \sigma}\left(D_V\frac{\partial c}{\partial \sigma}\right) + hS. \quad (3.35)$$

The equations above are discretised in an Arakawa-C staggered grid. The momentum conservation equations are solved using a finite-difference scheme, and the constituent transport equations are solved using a finite-volume scheme. Boundary conditions are specified for the bed (quadratic friction law) at the free surface (wave and wind stresses), at the lateral boundaries (water level, currents, and/or discharges), and at the coast as closed boundaries with free-slip conditions.

The flow and constituent transport equations need to be coupled with the wave model, which in this case consists of the SWAN spectral model (Booij et al. 1999). Spectral models solve a conservation equation for the wave action density, $N = E/\sigma_w$, where σ_w represents the intrinsic radian frequency of the waves and E represents the variance density:

$$\frac{\partial N}{\partial t} + \nabla_{x,y} \left[(\mathbf{c}_g + \mathbf{U}) N \right] + \frac{\partial}{\partial \theta} (c_\theta N) + \frac{\partial}{\partial \sigma_w} (c_{\sigma_w} N) = \frac{S_{tot}}{\sigma_w}. \qquad (3.36)$$

The velocity field $\mathbf{U}$ (m/s) is the 2D velocity field of the 3D current's top layer obtained in the (x, y, σ) hydrodynamic model or the depth-averaged velocity, depending on the model configuration. The depth-averaged velocity is defined as

$$\overline{\mathbf{U}}(x, y) = \frac{2k}{\sinh(2kh)} \int_{-h}^{0} \mathbf{U}(x, y, z) \cosh[2k(z + h)] dz. \qquad (3.37)$$

θ (°) is the direction of the wave field in space. $\mathbf{c}_g$ (m/s) is the wave group velocity. c_{σ_w} (rad/s^2) is the propagation velocity in direction and frequency space. S_{tot} includes several wave generation and dissipation mechanisms, including transfer of energy from wind to waves, energy dissipation due to white capping or bottom friction, and nonlinear energy transfer due to quadruplet (four-wave) and triadic (three-wave) interactions.

Several important considerations about the grid and the model convergence criteria need to be mentioned before we present the most relevant results. The bathymetry was reconstructed from USACE bathymetric measurements taken between 2002 and 2004. The bathymetric measurements were combined with LiDAR surveys of the intra- and supratidal shoal areas. In order to best capture the dynamics, the grid was divided into three curvilinear domains connected along their boundaries, covering each the ocean, the estuary, and the river sections, respectively. The domains follow the coastline contours in the relevant locations. The ocean domain has a coarse cell resolution of around 2 km on its seaward side, which is refined towards the MCR to around 200 m. The resolution at the MCR is between 100 and 200 m, in order to capture the dynamics around the jetties, in the navigation channels, and over the shoals. Within the Columbia estuary and the river approach, the resolution drops below 100 m and becomes finer as the river channels become narrower and shallower. The estuarine circulation is driven by residual currents resulting from the interactions of tidally driven and density-driven flows, and the secondary flows generated at the river bends; all these processes need to be captured accurately in the model. For the vertical discretization, a 20-layer model was used in the ocean and the estuarine domains, but the river domain was depth averaged, so the predictions of the model could be compared to tidal gauge data from a tide gauge located 86 km upstream from the MCR. A timestep of 15 s was used to ensure model stability of the domain with finest resolution—this timestep is applied to all three domains; this is the only option in domain decomposition models. Several sensitivity tests were carried out to ensure these space and time resolutions gave sufficiently accurate predictions, with positive outcomes. Several model calibration tests were also performed to ensure that the processes in all three domains were well captured and there was good agreement between the model predictions and the measurements—see Elias et al. (2012) for further details. It is important to note that the model does not include the response of the buoyant plume to the coastal forcings, and thus perfect agreement between the model and the measurements is not expected (Chawla et al. 2008). The agreement between the hydrodynamic model and observations was analysed using the 'index of agreement' or skill proposed by Willmott (1981):

$$\text{Skill} = 1 - \frac{\sum_{i=1}^{N} \left(X_{\text{mod}}^{i} - X_{\text{obs}}^{i}\right)^{2}}{\sum \left|X_{\text{mod}}^{i} - \overline{X_{\text{obs}}^{i}}\right| + \left|X_{\text{obs}}^{i} - \overline{X_{\text{mod}}^{i}}\right|}, \tag{3.38}$$

where X is the time series and $\overline{X}$ is its time average. The accuracy of the wave model, on the other hand, was quantified using several criteria for the significant wave height H_{m0}, the mean wave period T_{m01}, and the deep water steepness parameter H_{m0}/L_0. These criteria included the relative bias,

$$\text{rel.bias}_X = \frac{\text{BIAS}_X}{\frac{1}{N}\sum_{i=1}^{N} X_{\text{obs}}^{i}}, \tag{3.39}$$

the scatter index,

$$\text{SI}_X = \frac{\sqrt{\frac{1}{N}\sum_{i=1}^{N} \left(X_{\text{mod}}^{i} - X_{\text{obs}}^{i}\right)^{2}}}{\frac{1}{N}\sum_{i=1}^{N} X_{\text{obs}}^{i}}, \tag{3.40}$$

and the bias-corrected scatter index,

$$\text{BCSI}_X = \frac{\sqrt{\frac{1}{N}\sum_{i=1}^{N} \left(X_{\text{mod}}^{i} - X_{\text{obs}}^{i} - \text{BIAS}_X\right)^{2}}}{\frac{1}{N}\sum_{i=1}^{N} X_{\text{obs}}^{i}}, \tag{3.41}$$

where BIAS_X is defined as

$$\text{BIAS}_X = \frac{1}{N}\sum_{i=1}^{N} \left(X_{\text{mod}}^{i} - X_{\text{obs}}^{i}\right). \tag{3.42}$$

Some important results of the Columbia River case study are now presented, which highlight the importance of model calibration and validation through model parameter and boundary condition sensitivity analyses. First, initial simulations with constant values of the bottom roughness did not reproduce the tidal propagation dynamics in the upper estuary with good skill. A number of Manning and Chézy bottom roughness coefficient values were tested, and the values tested were different in each of the three domain-decomposition domains. An acceptable accuracy was obtained with a Chézy roughness coefficient of 61 $m^{1/2}$/s in the ocean domain, 55 $m^{1/2}$/s in the estuary domain, and 50 $m^{1/2}$/s in the river domain. Lower Chézy values upriver may be attributed to lower effective drag coefficients in the stratified part of the system (Giese & Jay 1989). The importance of the bathymetry and topography data was also highlighted during the model calibration tests, because increasing the grid resolution to represent better the major tidal flats and navigation channels upriver of Astoria, for example, improved the model predictions for water storage volume and tidal propagation in this location. These tests and calibrations are important not only to improve the model skill, but also to understand the dynamics of the system. In summary, the model calibrated with the measurements showed the following:

- The tidal forcing explained 99% of the ocean water level variance.

- The baroclinic tides started to be modified as they approached the river mouth, due to friction, density gradients, and wind and wave stresses. The tide corrections based on the data at the MGT stations varied from 0% to 5% in tidal amplitude and from 0° to 9° in tidal direction.

- The tidal waveshape and speed of propagation was further modified as the tide propagated upstream. Around Astoria, the tidal range increased due to the funnel shape of the lower estuary, while towards Skamokawa and Beaver, tidal modulation was reduced due to the opposing river flow and to bed friction.

- The accuracy of the salinity field structure predicted by the model depended, in turn, on the accuracy of the predicted tidal and subtidal flow and on the turbulence-closure parameterisation for mixing of momentum and salt.

- In general the skill values decreased upstream, due to coarse model parametrisations and to errors in the bathymetry.

- The water level and the along-channel velocity were dominated by the semidiurnal tide.

- Observed and modelled depth-averaged velocity magnitudes varied between 1.5 m/s during spring flood tides and more than 2.0 m/s during spring ebb tides.

- The model captured well the across-channel variations of the flow.

- Density gradients introduced a mean flow reversal at the bed and increased the mean and peak flow speeds in the upper part of the water column.

- Wind and waves had very small contributions on peak ebb and flood flow, with a 1%–10% increase of the residual outflow near the surface and similar decrease in inflow near the bed. This was due to the calm weather conditions encountered during the MGT experiment.

- In general, the larger wave heights observed during peak ebb flow tended to be underestimated by the wave model.

- The current velocity profiles in the MCR are highly stratified, and therefore, it is important to choose well the input current data to be used in Delft3D-SWAN for the wave–current interaction modelling. Here three different current velocity profile input options were tested, using either the top-layer current velocity, the depth-averaged velocity, or the wave-orbital-weighted currents. It was found that the wave-orbital-weighted current velocity profile yielded the best results.

Finally, it is important to note that the model was only calibrated for the conditions encountered during the MGT experiment. A calibration over a broader range of conditions should be undertaken. However, the model has been proved to work well under mild wave and wind forcing conditions, typical of the summer season.

3.4   Data-Driven Modelling

Data-driven modelling is an interdisciplinary field of science that focuses on statistical and pattern analysis of observations or model outputs. Data-driven methods may be divided into linear (Larson et al. 2003) and nonlinear techniques (Southgate et al. 2003). Examples of linear methods include bulk linear statistical methods, principal component analysis (PCA) and any variation of PCA, canonical correlation analysis (CCA), empirical orthogonal teleconnections (EOTs), or statistical clustering. Examples of nonlinear methods include extended empirical orthogonal function (EEOF), singular spectrum analysis (SSA), fractal analysis, wavelet analysis, or neural networks. Some of these methods are described in more detail in this section based on case study examples collected from the literature. Recent research has combined data-driven schemes with numerical models (Alvarez & Pan 2016) or with reduced-complexity approaches (Reeve et al. 2016) to extract coastal patterns and analyse their spatio-temporal behaviour. Empirical orthogonal function (EOF) methods were first applied to beach morphodynamics by Winant et al. (1975), who used monthly beach data from Torrey Pines Beach, California, collected over a period of 2 years. They found that most of the beach variability could be explained by the first three principal components, which corresponded, physically, to the mean beach profile, the onshore–offshore seasonal bar migration, and the low-tide terrace. This pioneering work with EOFs has been followed by several studies on, for example, barrier islands (Aranuvachapun & Johnson 1978), beach level variability through beach profile time sequences (Wijnberg & Terwindt 1995), and sandbank and coastal channel evolution (Reeve et al. 2008). Let ξ_l be the spatial coordinate and t_k the temporal coordinate and the bed level be defined as $g(\xi_l, t_k)$, with $1 \leq l \leq L$, $1 \leq k \leq K$. Note that any spatially 2D data can be expressed with one spatial coordinate, by reorganising the M-by-N matrix into a column vector with dimensions MN-by-1. EOF and PCA rely on the separation of the signal into spatial and temporal variabilities, so conceptually the goal of the method is to separate the signal as

$$g(\xi_l, t_k) = \sum_{p=1}^{L} \alpha_p c_p(t_k) e_p(\xi_l). \tag{3.43}$$

The normalisation factors depend on the dimensions of the domain, $\alpha_p = \sqrt{\lambda_p L K}$, and the functions e_p, the spatial eigenfunctions, with their corresponding the eigenvalues λ_p, are determined directly as the eigenfunctions of the square $L \times L$ correlation matrix of the data, $\mathbf{A}$,

$$\mathbf{A} e_p = \lambda_p e_p. \tag{3.44}$$

Note here that $\mathbf{A}$ is real and symmetric, it has L real eigenvalues, and its eigenvectors may be chosen as mutually orthonormal; that is,

$$\sum_{l=1}^{L} e_p(\xi_l) e_q(\xi_l) = \delta_{pq}, \tag{3.45}$$

where δ_{pq} is the Kronecker delta. The correlation matrix is calculated directly from the data. It has $L \times L$ elements a_{mn} of the form

$$a_{mn} = \sum_{K}^{k=1} \frac{1}{LK} g(\xi_n, t_k) g(\xi_m, t_k), \text{with } 1 \leq m, n \leq L. \tag{3.46}$$

The functions $c_p(t_k)$, the temporal eigenfunctions, satisfy

$$c_p(t_k) = \sum_{p=1}^{L} g(\xi_l, t_k) e_p(\xi_l), \tag{3.47}$$

with the normalisation factors now incorporated in c_p. From the properties of real symmetric matrices, one can deduce some of the properties of the data:

- The trace of **A** is equal to the mean-square value of the data or the energy.

- Each eigenvalue λ_p represents the relative contribution of mode p to the total variability.

- The matrix can be arranged so that λ_i are in decreasing order with increasing i, so that e_1 accounts for most of the mean-square value, e_2 for most of the remaining mean-square value, and so on.

- In general, the first five modes capture more than 90% of the total energy, unless the system has a lot of red noise (Vautard & Ghil 1989).

- The shape functions can be interpreted as 'modes' of variation, as in Fourier analysis.

- The number of maxima and minima in an eigenfunction increases with the order of the eigenfunction.

In the analysis above, we have only considered one spatial variable ξ_l, but a 2D space (x_i, y_i) can always be mapped onto a 1D array ξ_l, with a book-keeping method that helps arrange the eigenfunctions back into the original 2D space and plot them easily into contour maps for the interpretation of results. The dynamics that can be resolved with data-driven methods will depend very strongly on the spatial and temporal resolution of the data. It is essential to take those limitations into account from the moment the research is being planned. For example, in Reeve et al. (2008), the spatio-temporal resolution of the bathymetric data is sufficient to analyse the sandbank and channel evolution but not detailed enough to analyse the dynamics of ripples. In fact, due to the characteristics of coastal areas, it is often difficult to obtain appropriate datasets in order to apply the EOF method intensively. However, some data acquisition techniques, such as video cameras or satellite imagery, can help with the acquisition of beach data that lends itself well to the application of the EOF method (Fairley et al. 2009). EOF components allow describing changes occurring over the period covered by the data. However, EOF can be combined with different extrapolation techniques, in order to quantitatively assess the data variability beyond the observation period. As shown by Reeve et al. (2008), in most cases the extrapolation is limited to the EOF temporal component.

While EOF analysis is one of the most widely used linear data-driven methods, its nonlinear counterpart, the EEOF analysis or singular spectrum analysis (SSA), is less popular for beach data. However, it has been used to assess sandbar variability in the Baltic Sea (Różyński et al. 2001) and to analyse long-term trends in shoreline position at several beaches around the world (Southgate et al. 2003, Różyński 2005). SSA, which is described in detail by Elsner & Tsonis (1996), has been applied extensively to several oceanographic and meteorological time series (Ghil 2002). As with all data-driven techniques, SSA separates the data into a trend, some oscillatory components that can explain periodicity patterns at different time scales, and noise. With the multivariate version of SSA, the MSSA, the data matrix may include not only the bathymetry but also the forcing data. MSSA was applied, for example, to monthly time series of beach elevations, water levels, wave heights, and sea-level pressure differences associated to the North Atlantic Oscillation at Duck, North Carolina, to assess monthly to yearly sandbar migration cycles and the possible forcing mechanisms of such cycles (Magar et al. 2012).

Finally, current research has focused on the development of hybrid data-driven methods or on data-driven methods applied to numerical model outputs. Data-driven methods have also benefited in recent years from the availability of high-quality remote sensing imagery from satellites, the widespread use of stationary beach cameras, or imagery from drones. These data sources are more valuable when combined with *in-situ* monitoring of the forcings driving coastal change, such as waves, tides, river discharges, and storm surges. Reeve et al. (2016) argue that successful solutions will need to account for cause–effect dependencies between forcings and beach change observations, which can explain morphological tendencies and patterns at time scales of 10–100 years, and length scales of 10–100 km.

4

Future Modelling Applications

4.1 Climate Change: Storm Surge and Inundation Modelling

Here a climate change impact assessment is presented, together with an adaptation good practice (AGP) numerical modelling case study, developed for Bunbury, a small city in South Western Australia. The methodology is explained in detail by Fountain et al. (2010), and the aspects that are most relevant for regional modelling are summarised in this section. The purpose of the study was to develop a methodology with open source software that could be understood and implemented very easily by coastal authorities. The information published by the Australian National Climate Change Adaptation Research Facility (NCCARF) is publicly available—see www.nccarf.edu.au (NCCARF 2014), for this case study and for additional examples. The Bunbury case study is an example of coupled modelling applied to coastal inundation under different sea-level-rise scenarios, combined with worst-case storm tracks. The approach adopted involved integrating the outputs from Global Environmental Modelling Systems (GEMS) 2D Coastal Ocean Model (GCOM2D) regional-scale storm surge model with the free and open source hydrodynamic modelling software ANUGA to estimate inundation at the City of Bunbury resulting from a range of possible storm surge events. These models are used in conjunction with a Shoreface Translation Model (STM). The purpose of the study was to inform coastal practitioners and decision makers of the potential damage to coastal infrastructure under extreme events. For such studies to fulfil this purpose, it is necessary to analyse not only likely future climate trends, but also key climate change vulnerabilities, and different adaptation choices, to not only overcome barriers and minimize costs, but also identify, pursue, and fund priority research.

The methodology couples three component models that have been validated independently. So, one of the aims was to validate the coupled modelling approach. The three components consist of two hydrodynamic models that work best at different scales: the storm surge model GCOM2D (Hubbert et al. 1990, Hubbert & McInnes 1999); the open source, hydrodynamic, and hydraulic ANUGA model (see https://anuga.anu.edu.au/), developed by Zoppou & Roberts (1999) and initially released to the Open

Source community in 2006 through a Subversion (SVN) repository by Geoscience Australia; and the STM, developed by Cowell et al. (1992) at the University of Sydney. The model outputs may serve different purposes; for example, they may

- highlight the coastline areas that are most vulnerable to inundation;
- raise community awareness to coastal risks and coastal protection;
- inform town development policies, based on the risk of storm surge inundation; or
- develop emergency planning responses and flood protection infrastructure.

4.1.1   Storm Surge Model GCOM2D

GCOM2D solves the shallow-water momentum conservation equations in two (horizontal) dimensions. The model is driven by wind stress, atmospheric pressure gradients, astronomical tides, and quadratic bottom friction (Hubbert et al. 1990). At the air–sea interface, an atmospheric forcing derived from the lowest level of a limited-area, numerical weather prediction model is applied (Leslie et al. 1985) but with variations on surface roughness and with atmospheric instability taken into account through a boundary layer model. The nonlinear momentum conservation equations solved by GCOM2D are as follows:

$$\frac{\partial U}{\partial t} - fV = -mg\frac{\partial \eta}{\partial x} - \frac{m}{\rho_w}\frac{\partial P}{\partial x} - m\left(U\frac{\partial U}{\partial x} + V\frac{\partial U}{\partial y}\right) + \frac{1}{\rho_w H}\left(\tau_{sx} - \tau_{bx}\right),$$

$$(4.1)$$

$$\frac{\partial V}{\partial t} + fU = -mg\frac{\partial \eta}{\partial y} - \frac{m}{\rho_w}\frac{\partial P}{\partial y} - m\left(U\frac{\partial V}{\partial x} + V\frac{\partial V}{\partial y}\right) + \frac{1}{\rho_w H}\left(\tau_{sy} - \tau_{by}\right),$$

$$(4.2)$$

where the chosen map projection is taken into account explicitly through a map factor m and x and y are the horizontal Cartesian coordinates in the map projection's plane. Here (U, V) are the depth-integrated current velocity components, η is the free surface elevation, H the total water depth, f the Coriolis parameter, P the atmospheric sea surface pressure, $\vec{\tau}_s = (\tau_{sx}, \tau_{sy})$ the sea surface wind stress, and $\vec{\tau}_b = (\tau_{bx}, \tau_{by})$ the bottom shear stress. The continuity equation is of the form

$$\frac{\partial \eta}{\partial t} = -m^2\left\{\frac{\partial}{\partial x}\left(\frac{UH}{m}\right) + \frac{\partial}{\partial y}\left(\frac{VH}{m}\right)\right\}.$$

$$(4.3)$$

The surface wind stress and the bottom shear stresses are both determined using quadratic relationships. In the case of $\vec{\tau}_s$, we have

$$\tau_{sx} = C_D\rho_a\left(U_{10}^2 + V_{10}^2\right)^{1/2}U_{10},$$

$$(4.4)$$

$$\tau_{sy} = C_D\rho_a\left(U_{10}^2 + V_{10}^2\right)^{1/2}V_{10},$$

$$(4.5)$$

where (U_{10}, V_{10}) are the wind velocity components 10 m above sea level, ρ_a is the air density, and C_D is the sea surface drag coefficient. A number of formulations for C_D exist in the literature. For GCOM2D, Hubbert et al. (1990) apply the definitions adopted by Smith & Banke (1975), where C_D depends on a wind speed threshold. Below wind speeds of 25 m/s, C_D is defined as

$$C_D = \left[0.63 + 0.066 \left(U_{10}^2 + V_{10}^2\right)^{1/2}\right] \times 10^{-3}, \tag{4.6}$$

and above wind speeds of 25 m/s, the expression for C_D becomes

$$C_D = \left\{2.28 + 0.033 \left[\left(U_{10}^2 + V_{10}^2\right)^{1/2} - 25.0\right]\right\} \times 10^{-3}. \tag{4.7}$$

The bottom shear stress, on the other hand, is proportional to the current velocity,

$$\vec{\tau}_b = K\rho_w \left(U^2 + V^2\right)^{1/2} \vec{U}, \tag{4.8}$$

with $K = 2 \times 10^{-3}$ or 2.5×10^{-3}, depending on the model calibration (Hubbert et al. 1990, Flather 1984). Other bottom friction coefficient formulations, for example that of Mofjeld (1988), suggest that K should depend on the ratio H/z_0, where z_0 is the bottom roughness, which corresponds to the height above the seabed at which the no-slip condition is applied. In GCOM2D, K is a constant and is set to 2×10^{-3}.

The boundary and initial conditions are as follows. The normal component of the velocity vanishes at coastal boundaries. At open boundaries, the component of the current along the outward directed normal (U_n) depends on the water depth, the sea surface elevation, and the phase speed, $C_p = \sqrt{gH}$, of shallow-water waves propagating in the domain:

$$U_n = \frac{C_p}{H}\eta. \tag{4.9}$$

The sea surface elevation at the open boundaries will have two components, each of them driven by two different types of forces. One of them is the astronomical tide, which generates a sea surface elevation η^T. The other one is the sea surface elevation driven by meteorological effects, η^M. While η^T is fully predictable, η^M has a random nature. In the Bunbury case study, GCOM2D is forced with both η^M and η^T contributions to η. η^T can be determined very easily, using the TPXO8.1 tidal principal component data. η^M can also be determined very easily by setting it to a barometric displacement due to a deviation from an equilibrium, mean value $\overline{P} = 1013\text{hPa}$,

$$\eta^M = \frac{1}{\rho_w g} \left(\overline{P} - P\right). \tag{4.10}$$

These boundary conditions permit the passage of information between coarse-mesh to fine-mesh model boundaries and reduce to a minimum the generation of noise. The storm surge meteorological forcing used in the

Bunbury case study corresponds to the oceanographic and atmospheric conditions during the passage of tropical cyclone (TC) Alby along the southwestern Western Australia, between the 4th and 6th of April 1978 (see www.bom.gov.au/cyclone/history/wa/alby.shtml). The model used to determine the path and wind speeds of the hurricane is that of Holland (1980). Alby's path is shown in Figure 4.1. The equations and assumptions above constitute the model set-up, which needs to be solved numerically with an appropriate discretisation of the equations. In GCOM2D, the solutions are computed in a square Arakawa-C grid, and the equations are discretised using a finite-difference scheme, solved explicitly in a three-step integration. The first step considers the effects of the gravity wave and the Coriolis terms in the momentum conservation equations and solves the full continuity equation. It is worth noting that the continuity equation should not be split because the finite-difference scheme sometimes introduces some very large errors, in particular in areas with large local bathymetry gradients. The second step is an

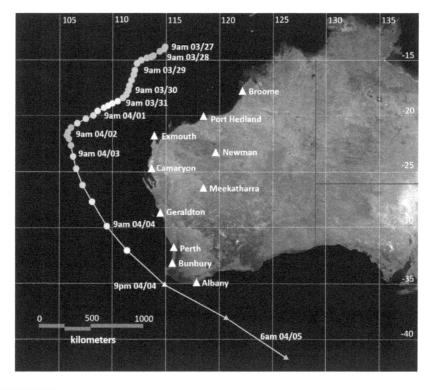

FIGURE 4.1
Track and the intensity of TC Alby (4–6 April 1978), according to the Saffir–Simpson scale. (Modified from https://en.wikipedia.org/wiki/Cyclone_Alby. [last modified: 29-01-2020].)

advective step, accounting for the remaining nonlinear terms. The final step is a physics step, accounting for surface-wind stress, bottom-friction stress and sea-level atmospheric pressure effects. Details on the numerical schemes adopted for each of the three explicit steps may be found in Hubbert et al. (1990).

4.1.2   Inundation Model ANUGA

ANUGA (Van Drie et al. 2011) is a local-scale, onshore inundation hazard model that has been successfully used for delta risk management, flood modelling, delta levee breach or dam break scenario modelling, and modelling of coastal processes, amongst others (see, e.g., Middelmann-Fernandes & Nielsen 2009). This is an open source, 2D model that focuses on storm hazard and potential flood impact modelling at local scales. It is an important first step in assessing flood impact under different scenarios and in reducing the costs of flooding to vulnerable communities. For coastal flooding analyses, the offshore boundary conditions for ANUGA need to be provided by a regional-scale storm surge model, such as GCOM2D. The extent of the modelling domain, on the other hand, needs to be defined according to previous storm inundation observations. In that respect, it is very important to consider a storm scenario which is still remembered by the local community, so that they can provide some qualitative information about the extent of the inundation and, ideally, about the damage caused by the storm to the inundated lands.

In contrast with GCOM2D, where the mesh is a Cartesian mesh, ANUGA is based on a triangular unstructured mesh, where the size and orientation of the mesh can be adjusted so the cells follow the orientation of the topobathy isolines or the boundaries of man-made structures. This is very important for the robustness of the model and also for the model to capture the inland propagation of the storm surge. The storm surge used as input data for ANUGA is generated by the GCOM2D model described previously. Specifically, ANUGA requires sea surface height and momentum at the boundaries, which ANUGA then computes within the domain cells at every timestep. Also, it is important to consider that the finest resolution of the model needs to be located in the most vulnerable regions in the domain. For the Bunbury case study, these regions correspond to the most valuable assets of the community, e.g., the commercial district, the community residential areas, the port and the marinas, amongst others. In such regions, the model developers assigned a maximum allowable area of 200 m^2 to the triangular mesh cells, corresponding to cell face lengths of around 20 m. The model simulation is started at $day\_time = 19780404\_120000$, that is, midday of the 4 April 1978, and is ended two full days later, at $day\_time = 19780406\_120000$. These 2 days are sufficient to capture the full wave dynamics and analyse the model predictions of the variations in water depth as the inundation moves inland, at selected synthetic observation stations. The advantage of time series analysis is that one can assess the properties of the inundating wave, for example, the number of peaks in the storm, as well as the characteristics of the largest peak. In order to determine the spatial extent of the inundation, however, it is necessary to generate inundation maps. Such maps need to be able to show the maximum distance travelled by the inundation storm waves.

ANUGA has a number of advantages and disadvantages, which are worth noting here. First it is a model that was originally developed for hydrological and hydraulic modelling. As such, it is very good for modelling hydraulic jumps, soliton-type waves, or tsunamis. Because the mesh is unstructured and the equations are discretised using a finite-volume scheme, ANUGA can reproduce the behaviour of complex flows, as well as transitions from subcritical to supercritical flow states (see https://anuga.anu.edu.au/). For inundation problems in particular, it is a very suitable model, as it is capable of simulating the wetting and drying process. It can also capture well the inundation process in complex terrains, such as in urban areas or around human infrastructure such as bridges, because of the flexibility of the unstructured mesh. Its main limitation, at least when the Bunbury case study was developed in 2009, was that it could not capture wave set-up processes. However, the storm surge model does so, as long as the coupling with the inundation model is sufficiently close to the beach; then the wave set-up component will, in principle, be within the boundary conditions transferring from the storm surge model into the inundation model. In ANUGA, a set of 2D shallow-water equations is also solved but using a methodology that allows it to capture shock waves. In other words, the method supports the existence of highly discontinuous solutions of the shallow-water equations. The method is the *weighted-average flux* (WAF) method, which was first presented by Toro (1989, as cited by Toro 1992).

4.1.3   Shoreface Recession Model

The 'Regional Storm Surge → Local Inundation' modelling process needs to be coupled with a shoreface evolution algorithm initialized from high-quality, digital elevation models (DEMs). The shoreface evolution algorithm determines the shoreface recession caused by the hydrodynamic and hydrological forcings. In this particular example, such DEMs are provided by the STM, which is a development of Cowell's group, from the University of Sydney. The method is based on behaviour-oriented equations coupled with an heuristic understanding of coastal patterns; a simplified methodology is described briefly by Magar (2008). This model bypasses small-scale coastal process uncertainties and provides a quantitative assessment of coastal evolution at spatial scales of hundreds to thousands of metres and temporal scales of tens to hundreds of years, which are the most relevant scales for climate change impacts and, therefore, critical for long-term coastal management. The STM simulates the evolution of mobile coastal sediment bodies, such as barrier dunes or mainland beaches, over solid, geological substrates. The response of the system will depend on the properties of the geological substrate and the active processes that the modeller considers. The properties of the geological substrate include the coastline configuration (the substrate's plan shape and slope) and the sediment type. The active processes that drive coastal change include sea-level rise, changes in sediment supply, the effects of storms, or changes in the shape of the sand body itself. Expert knowledge, whether based on uncodified expertise or machine learning algorithms, may also affect the model predictions and therefore the modellers' recommendations (after Cowell et al. 1995).

The STM is a cross-shore profile model and is regarded as a 1D profile because it is 1D in the horizontal. However, it is a 2D vertical (2DV) profile, as defined by Van der Werf et al. (2008), that describes the vertical evolution of the beach in the cross-shore direction. It takes into consideration the vertical evolution of the beach. The model considers two types of initial beach profile, a transgressive barrier beach profile (top of Figure 4.2) or an encroachment beach profile (bottom of Figure 4.2). The former corresponds to the case where the sediment may migrate inland through overwash and tidal inlet processes, while the latter corresponds to the case where erosional encroachment processes cause shoreface changes. For the encroachment case, the initial equilibrium beach profile below the encroachment point (or nick point) is based on the standard Bruun criterion (Bruun 1954), which has been shown to be adequate under a variety of conditions (Atkinson et al. 2018). If we consider the profile evolution over a period of years, rather than over storm events or seasons, then it has been found that many ocean-facing coastlines exhibit a concave curve which becomes more gently sloped with distance offshore. Bruun (1954) and later Dean (1991) showed that over yearly timescales, many

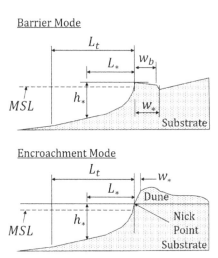

FIGURE 4.2
Sketches of the transgressive barrier and the encroachment beach profiles. (After Cowell et al. 1992.)

open coastlines will have a concave profile, with a profile depth h increasing seawards with distance x from the shoreline as

$$h = Ax^{2/3}, \tag{4.11}$$

where A, a constant related to the median grain size, d_{50} (in mm), is given as (Dean 1991)

$$A = 0.21d_{50}^{0.48}. \tag{4.12}$$

Thus, the equilibrium beach slope increases in steepness with increasing grain size.

The cross-shore model is applied for a series of transects, and the evolution of the beach along those transects has to be interpolated to obtain a map of beach recession. A similar procedure is used in work undertaken by the United States Geological Survey (USGS) along the coast of California (L. Erikson, Personal Communications, 2018). Recent modelling advances in 3D–2D–1D shallow-water modelling with a single modelling platform, such as Delft-FM, has led to some improvements in modelling approaches and numerical algorithms which, in principle, provide more accurate hydrodynamic predictions and morphodynamic evolution maps (Herdman et al. 2018). In fact, Delft-FM has been applied to levee breach scenarios with good results (Kerin et al. 2018), which demonstrates it is an acceptable alternative to ANUGA. Delft-FM is not an open source software, at least not at the time of writing, but using a single modelling platform may improve model development and user performance.

The models, the links between them, and their coupling with elevation maps for the topography and the bathymetry, are schematized in Figure 4.3, which synthesizes the descriptions of the modelling procedure given above. The flow chart of the procedure is summarised in the following section.

4.1.4   Model-Coupling, Hierarchical Procedure

The model-coupling, hierarchical procedure is 'model independent' and may apply to the coupled models described in this section, to the coupling of different Delft-FM modules, or to any other coupled modelling approach. So, from here on, the specific models used in the Bunbury case study are not mentioned. The coupled inundation modelling flow chart consists of the following reproduced from Fountain et al. (2010):

1. Build an offshore storm surge model.

   (a) Generate storm event using a vortex model.

       i. Set the storm path (geographic locations of the storm centre in time) from which the direction and speed of movement can be derived.
       ii. Set the central pressure at the eye of the hurricane.

      iii. Set the radius of maximum wind speed (i.e., the distance from storm centre to the band of strongest winds, in the eye wall)

      iv. Set the shape parameter B, defining the rate of change of wind speed with distance from the centre (along with the radius of maximum wind).

(b) Model the storm surge.

      i. Identify regions where inundation model will be applied.

      ii. Set the initial conditions. (We need three parameters: start day_time, initial sea level, seabed position.)

      iii. Set the boundary conditions (e.g., tides, surface wind speeds, sea-level pressure fields, and ocean–atmosphere heat flux model when necessary).

      iv. Perform a simulation over the set period of time.

      v. Determine wave height, sea level, and momentum time series at points on the boundary of the inundation model.

      vi. If modelling an historic event, calibrate the model by comparing the predictions with tide gauge measurements in the area.

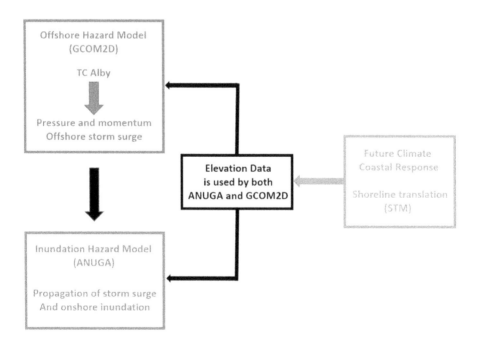

FIGURE 4.3
Synthesis of model coupling and information generated by each model. (After Fountain et al. 2010.)

2. Build an inundation model.

    (a) Identify regions where the model needs to be refined.

    (b) Set the initial conditions (day_time, initial sea level, seabed position).

    (c) Set the boundary conditions (outputs from storm surge model).

    (d) Perform a simulation over the set period of time.

    (e) Compute and analyse.

        i. Wave run-up (maximum elevation above mean sea level (MSL) reached by the wave).

        ii. Inundation distance (maximum distance from the coast reached by the wave).

        iii. Time series of inundation depth and horizontal momentum at strategic locations.

        iv. Verification or validation of these time series with observed data, if available.

3. Investigate the impact of existing storm surge protection infrastructure (e.g., storm gates, sluices, etc.).

4. Investigate the impact of different climate change inundation scenarios with a shoreface recession model.

    (a) Compute changes to the coastline caused by the sea-level rise induced by climate change, and return to 2b.

    (b) Change initial sea level in 2b; rerun inundation model.

4.1.5 Impact of Tropical Cyclones at Bunbury: TC Alby and Other Scenarios

The track of TC Alby shown in Figure 4.1 was used for the model coupling validation. Once the validation was successful, a number of 'worst-case' scenarios were tested. The worst-case scenario corresponded to a TC with same intensity and dimensions as TC Alby, but its path was changed until the track that caused most damage to Bunbury was identified. The results that showed the worst-track scenario at Bunbury were obtained with a TC track travelling almost parallel to the coastline on the northern side of Bunbury and making landfall near Bunbury. This track caused most damage in a 160 km stretch of coastline, extending around 80 km both to the north and to the south of Bunbury. This is due to the fact that TC Alby was travelling at around 80 km/h along this stretch, and therefore its impact all along this coastal area occurred within an hour, which for practical purposes can be assumed to be almost simultaneous.

    With the inundation model, it was possible to analyse the variations of the storm surge amplitude and flow speed over the simulation time. From these time series, the maximum inundation level and maximum flow speed maps could be generated at each location; these maxima are maxima over the simulation period. As for other regional models, one can analyse water depth variations at selected locations that either coincide with actual tidal gauges or play the role of synthetic water-level gauges. Then, one can analyse the peak surge data and determine their amplitude, time of occurrence, and number of peaks within the period of simulation. This information, together with the maps of inundation and flow speed maxima, is very useful for urban planning or emergency response activities.

A particular feature of Bunbury is the storm surge gate which helped reduce the inundation extent caused by the 1978 TC Alby event. The storm gate also acts as a storm water drainage for Bunbury. Therefore, if it is closed for too long, the embankment around the gate may be overtopped by inland water trying to reach the sea. Usually, high water levels occur when strong storms coincide with large astronomical tides, and with low pressures and strong winds, which cause rain and thus large demands of drainage runoff into the sea. A simulation was run to determine the recommended ocean water level at which the storm surge barrier should be closed. This simulation considered extreme rainfall and high ocean water levels for a sustained period.

Finally, the worst-track scenario was combined with different sea-level rise (SLR) scenarios, with sea level being artificially increased within the model from 0 to 0.4 m, 0.9, and 1.1 m. The following results were obtained:

- The model predictions in the validation case agreed well with water-level observations at tidal gauge locations, both with the GCOM2D–ANUGA coupled model and the inundation simulation run with GCOM2D alone.

- Simulation of inundation resulting from a 'worst-case' track using the current surface model and a sea-level rise between 0.4 and 0.9 m agreed well with the inundation observed in 1978 due to TC Alby.

- For all cases considered, the foredunes on the open coast were found to play a significant protection role against inundation.

- While the foredunes continue to exist in the potential future surface model, they are translated landwards and have increased in height.

The coupled model simulations that were run to assess the impact of increased storminess and sea-level rise on the coast gave satisfactory results for the Bunbury case study, and the model highlighted the importance of soft engineering structures, such as coastal dunes, in the protection of the coastline. The model, however, could still be improved further. For example, no attempt was made to couple the ocean inundation model with a coincident rainfall event or with riverine flooding. Also, ANUGA does not include wave set-up, which would increase the water-level predictions higher than those obtained at the time when the study was conducted (Fountain et al. 2010). Moreover, ANUGA cannot resolve vertical convection or 3D turbulence. Finally, the fluid is assumed to be inviscid, so the conservation equation does not include viscous terms (Geo 2019).

4.2   Coastal Erosion and Coastal Management

4.2.1   Coastal Protection Approaches

After the North Sea storm of 1953, the UK and the Netherlands invested many resources in building strong sea defences and levees to protect their coast. However, such defences are costly and difficult to maintain. Also, they create coastal squeeze problems, which have been exacerbated by climate change. In this section, we discuss some generalities about shoreline management and coastal protection, together with soft engineering methods that have become more important in recent years. In schematic format, there are five generic coastal protection policy options, as shown in Figure 4.4. These five options include the following:

1. Do nothing
2. Hold the line
3. Move seaward
4. Managed Realignment
5. Limited intervention.

The 'do nothing' approach would in many cases be the most appropriate approach, because significant stretches of coastline need no defending. An economic assessment of whether the benefits of defence outweigh the costs of construction needs to be carried out (Reeve et al. 2018), both within regional and national shoreline management plans, in order to optimize the use of resources and minimise the risks to valuable lives and assets. The level of

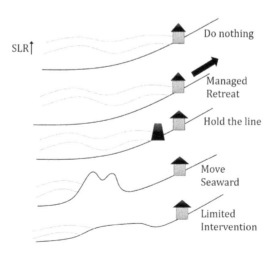

FIGURE 4.4
Schematic of coastal protection policy options.

risk can be quantified using numerical modelling procedures, such as those discussed in Section 4.1. However, while in many cases the inundation model in itself may be sufficient to assess the likelihood of inundation of some areas, depending on the storm surge scenario considered, in other cases this analysis may need to be complemented with computations for the probability of failure of sea defences.

The 'hold the line' approach is the classical coastal defence approach, involving the construction of defence infrastructure. This defence infrastructure may include one or more of the following:

1. Offshore reefs or breakwaters

2. Beach groynes

3. Rock armour or gabions

4. Seawalls

5. Revetments

6. Cliff stabilisers

7. Coastal barrages.

Offshore reefs are semisubmerged structures used to dissipate incoming wave energy and reduce wave erosion at the shore. On beaches protected by a series of offshore reefs, sediment builds up behind the reefs and, after a long time, a *tombolo* forms. A tombolo is an exposed sand spit joining the beach to the offshore reef. Many numerical modelling efforts of tombolo formation behind ridges have been undertaken, most recently by Alvarez & Pan (2016) and Razak & Nor (2018).

Beach groynes are fence-like, elongated structures built at right angles to the coast. Their purpose is to stop long-shore drift. Their upstream side suffers from sediment build-up, while their downstream side suffers from sediment erosion. If the groynes are sufficiently close together, however, there is no erosion on the downstream side of each groyne, and the groyne system acts as a wave energy buffer, thus protecting the beach behind it. Groynes and offshore reef systems have similar long-term caveats once tombolos have formed between the offshore reefs and the beach they are protecting from incoming waves. If tombolos form, the sediment is trapped upstream of the tombolo or the groyne, and erosion issues may manifest themselves at locations beyond the protected area. However, in some circumstances, this may be desirable, for example, when trying to control the amount of sediment reaching a reservoir; some authors have shown, with numerical models, that in this particular case a groyne field could be a very useful intervention (Mohammed 2017). Regional models are handy tools for morphodynamic analyses before and after the implementation of groyne systems, in particular at times when they may impact beach erosion processes (Gruwez et al. 2016).

Rock armour, gabions, or riprap are heaps of material used to armour coastal or river banks and protect them from erosion. Rock armour is found, commonly, at the toe of seawalls or bridge foundations to protect them from scour and from the damage that debris in the water may cause. The material absorbs the energy from the incoming sea waves or flooding waters, and the gaps between the material slow down the flow, reducing its ability to erode either the soil or the structures near the shoreline. There may be some regional modelling efforts regarding the evolution of beaches protected by rock armour units. However, the only reference found so far by the authors is a force and stress analysis on concrete armour units at local scales, performed with the model finite-discrete element (FEM-DEM) methods (Latham et al. 2014).

Seawalls are concrete and steel structures built parallel to the coast. They generally have smooth surfaces, but they may be enhanced with roughness elements to increase wave energy dissipation and protection against erosion or flooding (Hunt-Raby et al. 2010). By dissipating wave energy and preventing soil sliding into the sea, they fulfill their primary function as erosion protection structures. However, seawalls may also exacerbate erosion, in particular outside the regions they are protecting (Balaji et al. 2017), for the same reasons as beach groyne systems may cause significant problems. Models of scour at the toe of the structure provide information on the stability of seawalls, in particular for cases when no rock armour is present, to provide additional protection. Latest approaches to scour modelling include artificial neural networks and genetic programming (Pourzangbar et al. 2017). Some of the extreme events that cause the most damage to seawalls are, unsurprisingly, tsunami events. The results of many field studies are necessary for model verification and validation, such as that of Jayaratne et al. (2014) on tsunami-induced scour at the toe of sea defences. The studies have highlighted that the hydrodynamic parameters, the soil properties, and the characteristics of the coastal structure all play an essential role in the level of scour suffered after a particular tsunami event. The effects of a tsunami on soft engineering protections have also been analysed in recent work by Goff et al. (2009). Hydrodynamic and morphodynamic numerical models of posttsunami impacts on geomorphology are still lacking and worthy of further investigation. The modelling framework discussed in the context of tropical cyclones, discussed in Section 4.1, can be used here as well; because of its characteristics, the ANUGA model would be able to capture the inland-seaward propagation of a tsunami wave.

Cliffs can suffer erosion at the foot from wave action, tides and storm surge, overall erosion from the effect of the wind, sliding of the cliff, and changes of cliff slope due to waterlogging. Cliff stabilisation can be achieved by placing revetments at the foot of cliffs to protect them from geotechnical instabilities by dewatering the soil at the top of the cliff to reduce waterlogging and the probability of sliding or collapsing of the cliff (Mangor et al. 2017) or by artificial smoothing of the slopes (as has been done, for instance, in Point Grey, see https://web.viu.ca/earle/pt-grey/gvrd-pg-document-full.pdf). Some cliffs may not need protection when they are composed of a noncohesive material and rocks; the cliff slope will naturally erode slowly, and the cliff material falling at the foot of the cliff will provide some natural protection against storm waves. Engineering works to protect a cliff are more necessary when the cliff material consists of a mixture of sand, clay, silt, and rock. Different aspects of coastal cliff modelling are addressed by Castedo et al. (2017) and references therein.

The 'move seaward' approach is similar to the 'hold the line' approach while at the same time gaining land over the sea. This approach requires intensive beach renourishment campaigns and coastal defence schemes to maintain these beach renourishments in place. The 'move seaward' approach can be illustrated by a case study, involving climate change mitigation strategies adopted by New York City (NYC). Managed retreat is discussed in detail in a separate section, as it has been a fundamental coastal management approach during the 21st century. The limited intervention approach, as its name implies, refers to a coastal defence strategy, consisting of man-made structures or vegetation that induces wave energy dissipation and protects the coast against sea-level rise.

The impact of extreme events will differ with the type of coastal environment and with the level and type of coastal protection at the site of interest. Some of the most vulnerable regions are deltaic coastal systems, where sea-level rise relative to land level is much higher than the global average (Lowe et al. 2010). Figure 4.5 shows a 'critically eroded' coastline where the erosion threatens human and natural assets, such as infrastructure, habitats, or

FIGURE 4.5
Cliff erosion at Ponte Vedra beach, picture by Peter Willott for the St. Augustine Record. (Reproduced with permission from Martin 2018.)

important cultural resources. It illustrates soft cliff erosion in South Ponte-Vedra Vilano (SPVV) Beach, Florida, USA, in 2017, after hurricane Irma passed through the area. Some beaches along this coastline can lose up to 9 m of dune over a few high-tide cycles during severe erosion events, resulting in dramatic loss of land and, in some cases, condemnation of the structural assets. A 2008 study on the rate of erosion at SPVV showed it accelerated by a factor of 20 in the period between 2003 and 2008, compared to the erosion rate of the baseline period, i.e., between 1972 and 2003 (South Ponte Vedra-Vilano Beach Preservation Association 2015). This rate of erosion is due not only to climate change but also to site-specific factors, particularly to dredging campaigns at the nearby St. Augustine Inlet, which have disrupted the erosion–accretion cycle at SPVV. An option for improving the conditions at SPVV is to plan beach renourishment campaigns, increasing the width of the natural beach and providing natural wave energy dissipation habitats through managed realignment interventions, thus creating beach units where the natural erosion–accretion cycle can compensate for sediment losses through dredging. However, such beach renourishments are costly, and because SPVV is a private beach, home owners would have to cover half of the cost of the renourishments.

4.2.2   Managed Realignment and Shoreline Management

4.2.2.1   General Background

As for coastal environments, managed realignment sites will suffer from climate change impacts in a number of ways. Global warming will increase extreme event intensities and reduce return periods in ways that are not yet fully understood. This, combined with the water depth increasing and the sea reaching further inland than from sea-level rise alone, will mean that storm waves will reach further inland than predicted in control studies, damaging the coastal barriers and allowing waves to enter coastal lagoons, possibly all the way to the realigned defences. Generally speaking, rising sea levels will lead to low-lying land inundation, beach erosion, and intensified flooding. According to the Federal Emergency Management Agency, shore erosion is highly dependent on coastal exposure and beach setting, and it increases coastal vulnerability by removing the beaches and dunes that would otherwise protect coastal property from storm waves (FEMA 2015).

As an example of the impact of climate change on the coast, a study by Lowe et al. (2001) determined the frequency of extreme water level (see Figure 4.6), expressed as return period, from a storm surge model for present-day conditions (control) and the projected climate to around 2100, resulting from MSL rise and changes in meteorological forcing, for Immingham, a well-developed port on the low-lying east coast of England, on the Humber estuary. The water level is relative to the sum of present-day MSL and the tide at the time of the surge. The study is based on multidecadal integrations of the Hadley Centre regional climate model for the present climate and that at the

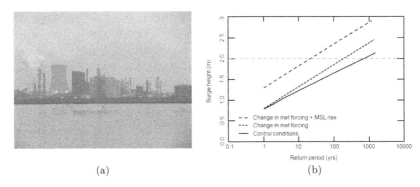

(a)   (b)

FIGURE 4.6
(a) Immingham port in the Humber estuary, UK (© Krzysztof, Google Guides) and (b) its surge height projections. (After Lowe et al. 2001.)

end of the 21st century, and it shows statistically significant changes of up to 0.2 m in 5-year extremes in the English Channel. Differences between these various results relate to the length of model integration and to systematic uncertainty in the modelling of both the atmospheric forcing and the ocean response. The red dashed line illustrates the return period shift, due to meteorological forcing and sea-level rise, of a surge with height of 2 m. The control curve shows this surge has a return period of 1,000 years. However, with meteorological forcing changes taken into account, the return period becomes 200 years, and with sea-level rise, it becomes ∼25 years. The drastic changes in return period associated with climate change effects on extreme events relate to the generalised extreme value (GEV) distribution satisfied by these extreme events.

Coastal management and coastal engineering used to be synonymous concepts. However, their modern definitions have diverged (Kamphuis 2000). This divergence is due, primarily, to an increased interest in the environment of our present society and to the implementation of environmentally friendly coastal protection schemes. Hence, while coastal engineering still focuses on coastal design and construction that provides safety, defence, and transportation, coastal management now also encompasses additional activities, such as stakeholder involvement, management of the coastal ecosystem, water quality and dispersion of pollutants, or retreat to more defensible shoreline positions, to name but a few.

One of the most important current issues that coastal managers have to address is coastal squeeze. Coastal squeeze refers to loss of intertidal habitats, due to coastal structures that disrupt inland habitat migration when water levels rise. This happens, for example, in the 'hold the line' approach to coastal protection, shown in Figure 4.4. Together with land reclamation, coastal squeeze constitutes one of the main causes of loss of habitats in human-modified coastlines (Doody 2013). The term coastal squeeze first appeared in Britain in the late 1970s and early 1980s, when the sustainability of the then land reclamation practices was questioned (Doody 2013). By reversing coastal squeeze, natural habitats could be recreated. Natural beach habitats are dynamic environments that absorb wind, tide, and wave energy and provide a barrier to flooding, as a wider coastal zone is likely to be more resilient to perturbations. This recognition prompted the start of a number of land restoration projects in Britain, based on techniques adopted in Germany that proved unsuccessful, demonstrating the site specificity of erosion and accretion processes. Therefore, the British government adopted a different approach, giving rise in the 1990s to regional shoreline management plans (SMPs), and proposed four possible response strategies to coastal change (DEFRA 2006): advance the existing defence line; hold the existing defence line; allow the shoreline to move backwards or forwards through managed realignment; and the 'do nothing' scheme. These strategies, together with the 'limited intervention' strategy, were introduced and discussed in some detail earlier.

Up until the end of the 20th century, the most popular coastal management approaches were to advance the line or hold the line of defence. However, from the beginning of the 21st century, coastal policy shifted to accommodate habitat loss due to coastal squeeze and sea-level rise. This was due, partially, to concerns relating to global warming and ageing coastal defence infrastructure, but a social awareness in relation to coastal management also had an influence. These three factors led to an increase in popularity of the managed realignment approach for coastal defence. Moreover, the EU Habitats directive, adopted in 1992 to protect natural habitats and migratory species in response to the Bern Convention (Council of Europe 1979), provided a clear opportunity for managed realignment to replace coastal hard engineering schemes and help generate new salt marsh habitats. Coastal squeeze consists of a loss of intertidal habitats between the sea and an existing line of coastal defence due to sea-level rise (Esteves 2014). Because of such defence, the intertidal habitats cannot move landwards, as sea level rises, the waves start to reach the defences as well, and the beach fronting the defence becomes more vulnerable to incoming waves. Thus, the intertidal habitat is lost, and the beach starts to erode more rapidly than before. One way of correcting this is to artificially breach or remove the defences in some locations, so that the water can flow freely inland and restore the intertidal habitat behind the defence (Leggett et al. 2004). This is only possible if the economic or cultural assets are no longer worthy of protection or have relocated elsewhere. Key to managed realignment is the concept of restoration of coastal habitats that can dynamically adjust to changing environmental conditions, so they can continue to provide ecosystem services to society, as defined by the Millennium Ecosystem Assessment (2005).

In order to increase the success of a managed realignment intervention, different flood defence redesign scenarios can be modelled, thus reducing negative outcomes and increasing the benefits of the intervention. Different types and levels of redesign of the defences can be tested. This may involve deliberate defence breaching, defence abandonment, or relocation (Nicholls et al. 2007), where the relocation may mean both shoreline advance or retreat (DEFRA 2006). There are different reasons why a shoreline realignment scheme may be desirable, including some of the following, in order to

- avoid negative impacts of artificially fixed coastlines
- work with natural processes and allow coastlines to evolve dynamically
- restore the environment to its natural functioning
- address environmental and economic problems caused by hard engineering
- compensate or offset intertidal habitat loss
- provide other ecosystem services such as carbon sequestration, amenity, or health value.

Managed realignment plays an essential role in SMPs. SMPs need to be developed in order to identify what, how, where, and how large managed realignment projects should be. Managed realignment requires, as well, an understanding of coastal processes, coastal defence needs, environmental considerations, planning issues, and current and future land use, all at an adequate level of detail. Assessing risks is also an important part of the managed realignment appraisal processes to ensure that the decisions taken are effective and based on an awareness of the consequences. This would also ensure that the appropriate measures needed to address these consequences are taken. This is paramount, particularly when considering that 77% of the total, global economic value produced by the world's ecosystems is generated at the coast (Martínez et al. 2007).

Most managed realignment schemes have been undertaken at latitudes where such schemes naturally promote the creation of salt marshes, as in Northwest Europe and North America. Salt marshes are well-known wave energy attenuators, so at the level of the new shoreline or set of defences, the system behaves simply as a tidal lagoon, in which the sediments are predominantly composed of silts and muds (French 1997). Both wave attenuation and the sediment type in the environment help reduce flooding and erosion, both landwards and seawards of the old set of defences. It is noteworthy that creation of other intertidal habitats should be possible. For example, one could envisage the creation through managed realignment of mangrove forests. Protection of urban areas through managed realignment could include the construction of floating gardens. Depending on the local needs, one can imagine many possible realignment schemes. The best scheme will, generally, be the one promoting the socio-economic, environmental, and legal sustainability of coastal erosion and flood risk management through the recreation of natural habitats and defences that can evolve dynamically with changing environmental conditions.

Managed realignment schemes should benefit from modelling and site monitoring before their implementation to help determine the effects of different implementation strategies. In this way, the optimal strategy (the one maximising the benefits and minimising negative impacts) can be adequately identified and thoroughly understood by the local communities. Most of the schemes implemented to date have some of the following characteristics and provide some of the following types of protection (Morris 2012):

- Salt marsh re-creation in historically low-wave energy sites

- Sheltering from wind-driven waves

- Construction of the site relatively high up in comparison to the mean height of the tidal prism

- Quick gains in ground elevation of managed realignment schemes, because they are natural sediment sinks

- High levels of sedimentation and vegetation development (dependent on the breach shape and size)

- Sediment remobilisation through tidal scour, wind-driven waves, and suspended sediment (SS) transport.

4.2.2.2   Coastal Resilience and Regional Modelling Opportunities

Early engagement with communities and stakeholders is essential in order to avoid public opposition which may delay, or even prevent, managed realignment projects. The success of shoreline management projects depends on several factors, including the funding source and the amount of funding, who will have the management responsibilities on the use of the shoreline and natural resources in that region or the potential degree of loss of existing assets, which could include loss of fishing grounds, access rights, cultural heritage, or damage to natural scenery (Reisinger et al. 2014, 2015, Alexander et al. 2012). This loss of assets could affect other economic activities, such as tourism. Therefore, a crucial aspect in shoreline management plans is to engage with the managers of marine protected areas (Cicin-Sain & Belfiore 2005).

Several countries have adopted managed realignment approaches to protect their coastlines, even for highly developed coastal areas. It may seem counterintuitive to adopt soft engineering approaches in highly vulnerable regions. However, such schemes recognize the need to implement dynamically adaptive measures, which are less costly and more effective in the long term, that recognize the uncertainties around sea-level rise and coastal storminess projections (Kelly 2015). Some managed realignment schemes implemented in coastal cities are analysed in what follows based on the information and references provided by Glavovic et al. (2015), Esteves (2014), and Ortega-Rubio (2018). As New York is one of the most emblematic coastal cities in the world, an NYC case study is discussed initially (Solecki et al. 2015). As with any city, numerical modelling tools may inform NYC architects and planners about climate risk and help them assess the benefits and costs of different coastal adaptation strategies.

NYC Case Study

The New York–Newark–Jersey City metropolitan area is the largest in the US, with an estimated population of almost 20 million by 2018. NYC has a population density of 27,000 people per square mile, or 10,400 people per km^2, the largest density not only in the US but also in the American continent, given that the population density within Greater Mexico City averages 25,300 people per square mile. The New York–New Jersey (NY–NJ) Upper Bay is a large estuarine harbour, fed by the Hudson river and connected to the Atlantic. The connection to the Atlantic occurs through the narrows linking the NY–NJ Upper Bay with the NY–NJ Lower Bay, and the Long Island Sound, adjacent to the boroughs of Bronx and (Upper) Manhattan. The NY–NJ Upper Bay has a surface area of ∼52 km^2, and it measures nearly six km from Saint George (in Staten Island) to Governors Island. This vast body of water is surrounded by the dense urban development of NYC. Adjacent to the NY–NJ Upper Bay are the three boroughs of (Lower) Manhattan, Brooklyn, and Staten Island (see Figure 4.7).

A number of researchers have analysed the potential impacts of climate change and hurricane impacts on coastal flooding in NYC. Nordenson et al.

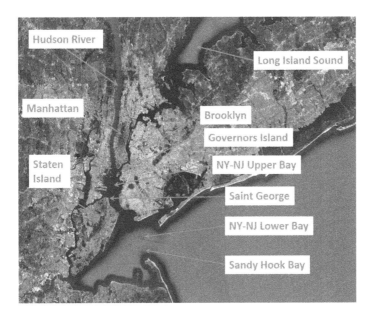

FIGURE 4.7
Main geographical features of NYC. (Produced with ©Google Earth.)

(2010), based on maps of flood projections using geographical information systems (GISs) of NY–NJ Upper Bay, assessed the size of the inundated regions due to coastal flooding for a wide range of sea-level rise and storm inundation scenarios: from one foot sea-level increase, which is the conservation projection, to a twenty-eight foot sea-level increase, corresponding to the level of flooding expected from a type 4 Hurricane, as predicted by the National Hurricane Center (NHC) 'Sea, Lake, and Overland Surges from Hurricanes' (SLOSH) computer model. SLOSH takes into account a number of variables for storms moving in different directions and varying in strength from category 1 to category 4. The model then calculates surge levels for a worst-case scenario (above NAVD88 bare earth ground elevation) for each storm category entering the New York Bight. A potential storm flooding map for NYC, under different hurricane categories, is shown in Figure 4.8. The different tones of shading go from hurricane category 1 to hurricane category 4, from the coast and landwards. With a category 4 hurricane, the inundation depths in all the locations mentioned in Figure 4.8 would be above 6 m. The NHC model takes into account the uncertainties in the tropical cyclone trajectories, intensities, the radius of maximum winds, as well as landfall location or angle of landfall. The flooding maps consider the effect of the storm surge in rivers, estuaries, and any area affected by the astronomical tides. The flooding maps are based on relatively detailed land elevation and coastal bathymetric data, but in complex regions, there may be some inaccuracies and thus a certain degree of uncertainty at a local level.

Nordenson et al. (2010) proposed several soft engineering interventions all along the shoreline of NY–NJ Upper Bay, as well as several floating gardens,

FIGURE 4.8
Potential storm surge flooding map for NYC. (From New York City Government 2019.)

and proposed a renaming of the Upper the Bay as 'Palisade Bay', referring to the ability of the bay to protect from incoming, extreme events. Although they do not have the capacity, with their tools, to analyse how their interventions reduce coastal flooding and its impacts, in the region, their approach follows the managed realignment conceptual framework by providing a solution that can adapt dynamically to sea-level rise. However, megacities are vulnerable to other climate change risks, such as heat waves and inland flooding (Depietri et al. 2018). Therefore, a multihazard analysis that takes into account all these risks is recommended, even if, at least in the case of NYC, coastal flooding has been shown to dominate over other climate change hazards (Depietri et al. 2018); and climate change adaptation strategies in the coastal areas of NYC should be prioritized.

San Pablo Bay (SPB) Case Study

The San Francisco Bay Delta watershed contains the only inland Delta in the world, covers around 200,000 km^2, and includes the largest estuary on the west coast of the American continent (US EPA n.d.). San Pablo Bay (SPB), in the northwest part of San Francisco Bay (see Figure 4.9), underwent considerable morphologic development in response to variations in fluvial sediment load and discharge associated with a period gold mining and later damming in the watershed (Elmilady et al. 2018). The SPB case study is a long-term analysis of morphological evolution under SLR using process-based modelling. SPB is in the northern part of San Francisco Bay. This is a unique region for morpho-dynamics studies, because bathymetric surveys have been undertaken almost

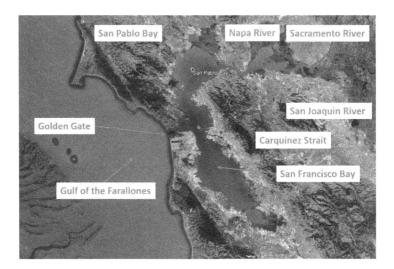

FIGURE 4.9
Map of the San Francisco Bay and SPB areas. (Produced with ©Google Earth.)

every 30 years since 1856. The last available survey was taken in 1983. Here the focus is on the work of Elmilady et al. (2018), who developed one of the first long-term process-based models of SPB, using the Delft3D model suite, to compute the hydrodynamics, sediment dynamics, and morphodynamics within the modelled domain. The model provides a hindcast over 127 years, for the period 1856–1983, when bathymetric data is available, and a forecast for 120 years, from 1983 to 2103. Elmilady et al. (2018) provide detailed information on the oceanographic and meteorological conditions that affect the dynamics of SPB, and the San Francisco Bay Delta watershed in general, which include the following.

- Waves and wind: Several authors have found that locally generated wind waves play an important role in both the shallow and the deep regions of SPB, where they promote sediment resuspension and therefore increase the suspended sediment concentrations (SSCs) (Bever & MacWilliams 2013). These waves are an important forcing that can reshape the seafloor in SPB, in particular in the muddy shoals, as reported by Ganju et al. (2009), Van der Wegen & Jaffe (2014), and Van der Wegen et al. (2016). Bed shear stress induced by the waves can cause important vertical gradients of the SSC and sediment/induced stratification (MacVean & Lacy 2014). Averaged wind speeds over a 7-year period are around 4.5 m/s (computed from wind data statistics retrieved from Point Davis weather station, located close to the junction between SPB and Carquinez Strait, and stored in Windfinder (2019)), with maximum sustained speeds of 24 m/s (Weather2 2019). The dominant wind directions are from the west and the southwest (wet and dry season, respectively) with little wind during the night.

- Tides: Bever & MacWilliams (2013) have reported tidal ranges between 1.0 and 2.5 m in SPB during neap tides and spring tides, respectively. The tides in SPB originate in the Pacific Ocean and enter the Bay as a mixed, semidiurnal, progressive tide. It can propagate quite far into the Sacramento and San Joaquin rivers during periods of low river discharge but is dampened out at the mouths of the Sacramento and San Joaquin river reaches otherwise (Kimmerer 2004).

- River inflow: The Sacramento and the San Joaquin rivers provide most of the freshwater discharge into SPB, with a contribution of 80% and 10%–15%, respectively, of the discharge. The remaining discharge comes from small tributaries, such as the Napa or the Petaluma rivers, or the Sonoma Creek. There is a very strong annual cyclic pattern of river discharge, with discharges of 16,000 m^3/s and of 300 m^3/s in the wet and dry seasons, respectively (Kimmerer 2004, Barnard et al. 2013).

- Morphology: SPB has an area of ∼300 km^2 and comprises a channel–shoal system of northern and southern shoals, where the water depth is below the 5 m MLLW (mean lower low water) line, and a deep main channel with a depth of 12 m MLLW or more (Schoellhamer et al. n.d.). The northern and southern shoals cover an area of about 190 and 43 km^2, respectively (Bever & MacWilliams 2013), corresponding to almost 80% of SPB's area. The Bay's margins comprise large intertidal mudflats, which are bound by salt marshes in some locations.

- Sediment characteristics: In SPB, 90% of the shallow tidal flats are covered with fine cohesive sediments (mud), while the channel is covered with sand, with some dynamic patches of mud (Locke 1971). This separation of sediment types is attributed to the high tidal currents in the channels and low flow velocities in the shallow zones. The low velocities allow for mud deposition, while high tidal velocities cause tidal scouring in the channels.

- Sediment load: The sediment to SPB is supplied mostly by the Sacramento–San Joaquin Delta, but since the 1950s, there has been a reduction in sediment supplies to the Bay due to a decrease in hydraulic mining, a decrease in urbanisation sediment pulses, sediment trapping behind dams and in flood bypasses, and bank protection (Moftakhari et al. 2015). This has resulted in erosion and loss of bed material in SPB (Schoellhamer 2011).

- Regional SLR: Council (2012) forecast an SLR of 4–30 cm by 2030, 12–61 cm by 2050, and 43–167 cm by 2100 compared with 2000. These sea-level increases may be implemented in different numerical model scenarios, and the effect of SLR on the hydrodynamics, the sediment dynamics, and the morphology can thus be assessed in detail, as discussed by Elmilady et al. (2018).

The environmental conditions and their changes, either because of climate change or because of direct anthropogenic interventions, clearly are changing the dynamics of the system. So the questions several researchers have asked is: by how much will such changes affect the dynamics of SPB in the future? For example, Kimmerer (2004) and Barnard et al. (2013) have pointed out that the mudflats and the salt marshes should undergo a restoration program in order to avoid a decline in their area associated with the rise of the MSL in SPB. If such a rise occurs, then the levee systems will be exposed to wave action, increased tidal action, and increases in SS discharges, which could result in severe decreases of SSC and erosion in the Delta region (Achete et al. 2017). However, the difficulties associated with modelling long periods in a reasonable time and with available computer power have only been solved by using modelling acceleration techniques (Elmilady et al. 2018) or schematic or representative forcing conditions. Schematic conditions consist of idealised forcings, while representative conditions are conditions for which the observed erosional or depositional pattern agrees with observations. In Elmilady et al. (2018), long-term modelling is achieved by adjusting the morphological factor to adequate values, either for the depositional or the erosional period.

The information available for SPB shows a very clear separation between a depositional (1856–1951) and an erosional (from 1951 onwards) period. In order to analyse the effect of SLR on the intertidal mudflats, Elmilady et al. (2018) set up a 250-year model with three SLR scenarios of 42, 84, and 167 cm by end of the 21st century, as explained earlier. The model design was based on previous 30-year modelling efforts by Van der Wegen et al. (2011) and Van der Wegen & Jaffe (2013) for depositional and erosional periods of SPB, respectively, with some small modifications. Van der Wegen et al. (2010) showed how a sediment dynamics module that allows for sediment fraction redistribution over the seabed, but with no morphodynamic changes, improves morphodynamic prediction when used as a premorphodynamic modelling 'spin-up' approach, as long as it then allows for a natural development of the morphology. Van der Wegen et al. (2010) also assessed the sensitivity of the morphodynamic model to the choice of morphological factor, the active layer thickness, and wind waves. The active layer is the layer of sediment that can be mobilised by the flow, and its thickness can vary from a few centimetres to a few metres, depending on the sediment composition and the environmental forcing conditions. Van der Wegen et al. (2011) used the process-based model combined with the 'spin-up approach' of Van der Wegen et al. (2010) to analyse the dynamics of SPB during the depositional period between 1856 and 1887, when a sediment pulse due to hydraulic mining increased significantly the sediment loads into SPB. Van der Wegen et al. (2011) carried out a thorough sensitivity analysis to assess the relative importance of different processes and the required accuracy of the process descriptions and model parameter values. The model includes a river discharge forcing from the Carquinez strait into SPB that is highly schematised. Van der Wegen & Jaffe (2013), on the other hand, constructed an ensemble of input parameters and schematic forcing conditions and assessed for each model the model skill or the model performance, or both, in predicting the observed depositional patterns for a 30-year model simulation during the SPB depositional period extending from 1856 to 1887 or the erosional patterns for a 30-year model simulation during the SPB erosional period extending from 1951 to 1983. The motivation behind these model skill and model performance analyses is that model predictions can only be used for estuarine management when the model uncertainties have been quantified in detail (Vreugdenhil 2006), in particular when making management decisions based on model forecasts. In the work of Van der Wegen and collaborators, long-term modelling is achieved by adjusting the morphological factor, MF, to appropriate values, according to the modelling objectives. They adopt a holistic modelling approach, as defined in Seminara & Blondeaux (2001), by involving detailed description of physical processes and using tools capable of modelling the complete nature of the system (Van der Wegen et al. 2011). Moreover, they cite previous work by

Van der Wegen & Roelvink (2008), who showed that 'one month of calculations with a morphological factor of 400 leads to similar results as 40 months of calculations with a morphological factor of ten' (Van der Wegen et al. 2010).

The model set-up used by Van der Wegen et al. (2011) has the following characteristics:

- 3D, hydrostatic, shallow-water model Delft3D set-up with a curvilinear grid in the horizontal coordinates and a σ-layer (terrain-following) grid in the vertical coordinate.

- $k - \epsilon$ turbulence-closure model is used to model turbulence at subgrid scales.

- Allows for salt and freshwater density variations.

- Allows for variations in bed composition and specification. This means that bed layers with different percentages of mud and sand can be defined in the model and that spatial variations of the critical shear stress have been implemented.

- Wave forcing is included through coupling of the Delft3D-flow module with SWAN.

- A schematized wind field is applied over the water surface of the bay.

- The water levels and flow velocities are calculated every minute, but the wave coupling and wave-induced shear stresses are computed every hour only.

- The grid cells have a size of 100 m by 150 m, and the timestep selected for model runs is 2 min.

With this set-up, a quad-core PC with 3 GHz can run a 4-month hydrodynamic simulation in around 36 h. The model has 15 σ levels, allowing for density currents and waves. The σ layers are close together near the bed and near the water surface in order to compute the waves and also the surface and the bed shear stresses with more precision. The model is run for a schematised year with highly schematised tide, wind, and wave conditions, as well as sediment inputs.

Further consideration of the characteristics of the morphodynamics indicate that it is strongly influenced by a yearly cycle consisting of a short wet period and a long dry period. Ganju et al. (2008) describe how to derive a daily discharge and a sediment load distribution for the Sacramento River and San Joaquin River that, when integrated in time, agree with decadal sediment load estimates. This defines the *yearly morphological hydrograph* concept, which, when combined with a process-based model, captures the high variability of the system at decadal timescales but at a reduced computational effort. The sediment load time series and the morphological hydrograph concept can be used as landward boundary conditions for hindcasting morphodynamic simulations for studies of contaminant deposition, for marsh accretion, and for climate change impacts on the San Francisco Bay. By assigning a contaminant concentration to the sediment load estimates, the mass of sediment-associated contaminants exported from the watershed may be estimated. These sediment load estimates can then be combined with observations of marsh accretion, over decadal timescales, to calculate potential trapping efficiencies of estuarine embayments and adjacent marshes. Finally, in view of future scenarios of climate and land use change, reconstructed time series such as this provide bounds on historical conditions in watershed–estuary systems.

4.3   The (First) Sand Engine, the Netherlands

Beach renourishment is a technique to either restore a beach to its pre-eroded state or artificially modify the shoreline or shallow areas for engineering projects. Beach renourishment is a method of coastal management which was first implemented in the early 1900s but is still very popular because it preserves the natural beauty of the area. However, it needs to be an ongoing process, involving periodic dumping of sediment to maintain the beach. This is due to the dynamic nature of the beach material, which will be transported by waves, tides, and winds and will change the shape of the beach as a response to those forcings. Given its increase in popularity, there is also an increase of case studies in the literature, as well as a development of good practices for the planning, design, implementation, and monitoring of beach renourishment projects. Numerical modelling schemes are beneficial for planning beach renourishment projects and also crucial tools that provide understanding of the underlying causes of beach erosion. This is achieved through understanding of the natural processes that have shaped the beach to its current state and assessing its evolution of the renourished beach if a renourishment scheme is implemented. Also, modelling tools can help to analyse different beach renourishment options and, in particular, help to evaluate the evolution of the renourishment scheme into the future.

Most of the restrictions imposed on beach renourishment schemes are practical restrictions, which reduce costs. For example, it is very important to source the sediment for the scheme at sites within a reasonable distance from the project location. Once where and why a beach renourishment project is necessary is determined, then where and how sediment should be delivered to the shore needs to be planned, and the difficulties likely to be encountered during and after renourishment assessed. A preliminary investigation should include (Bird & Lewis 2015)

- the dimensions and morphology (with transverse profiles) of the beach and nearshore (including the breaker zone);
- the relationship of the beach to nearby cliffs, bluffs, reefs, river mouths, tidal inlets, and drains;

- grain size (modal, range) and shape (rounded, angular);
- proportions of sand, granules, pebbles, cobbles, shells, rock fragments, alien particles (brick, glass, earthenware);
- dominant minerals (quartz, feldspar, carbonates, olivine);
- evidence of source(s) of beach sediment (continuing or relict);
- wave regime, storm effects, seepage, run-off;
- tide ranges (neap, spring);
- evidence of alongshore drift (dominant direction or seasonally alternating);
- evidence of offshore–onshore movements of beach sediment;
- presence of microcliffs, berms, and wash-overs;
- evidence of history (maps, remote sensing, previous surveys);
- evidence of rates of erosion or accretion over a specified time;
- indications of cause(s) of beach erosion.

An emblematic beach renourishment project of the 21st century is the Sand Engine project. The Dutch coast in general is very vulnerable to climate change, but in particular, the west part of the Netherlands, near the sea, is below sea level and hence needs to be protected. To do so, the Dutch government undertakes a quinquennial programme of beach renourishment. This programme is very costly, and if not undertaken, the beaches would be eroded. So the Sand Engine project was proposed in order to assess the ability of alternative schemes to maintain the beach health in a more sustainable way. The objective of the project is to make the most of what is known as the *economy of scale*; that is, this single, large-scale, beach renourishment scheme, is expected to protect the beaches in the Delfland region for the next 20 years, avoiding the recurrent costs of the quinquennial programme (Sand Motor 2019). Because the Sand Engine project is the first of its kind, it is considered a pilot project and, therefore, requires a very close follow-up in order to assess whether it is being successful in achieving its objectives.

4.3.1 Description of Project and Field Observations

The Sand Engine pilot project, through a one-off, megascale renourishment scheme, led to the creation of a hooked peninsula. This changed the shape of the coastline and therefore modified the dynamics of the nearshore currents, waves, and sand transport processes. Several scientists have studied this system from different perspectives since 2011. Their investigations focused not only on core science and engineering questions about the performance of the project, but also on ecological and social science questions regarding the impact of the scheme on local wildlife and on different human activities such as tourism, water sports, or fishing. Figure 4.10 shows an infographics with the different research areas and beach monitoring research activities undertaken since the start of the project and shows an almost symmetrical spread of the Sand Engine materials, with an observed beach accretion that is very mildly larger towards the North of the Peninsula than towards the South. The volume of sand used for the project was around 18.7 million m^3 of sand, which was deposited off the coast of the Den Haag and Westland municipalities (Buitenkamp et al. 2016). Wave buoys and current profilers and radars were used to measure currents, water levels, and wave heights and directions. A mast with eight cameras was set up to monitor the evolution of the Sand Engine through time. Benthic analyses were carried out in the coastal zone and on the beach together with regular bed measurements. Monitoring of the coastal dunes included assessment of the vegetation, bird surveys, geomorphological measurements, and sand and salt spray measurements. Groundwater levels were also measured. Studies analysing the social impact of the scheme included, for example, a survey of users' experiences and bathing safety indicators.

There were a number of questions to answer with this project. For example, if the scheme can replace the quinquennial beach renourishment program, then the sand needs to spread and protect the target area within 5 years. So, will the sand spread quickly enough for this to be achieved? Will the Sand Engine work as an artificial replacement of natural habitats in the Netherlands? How will

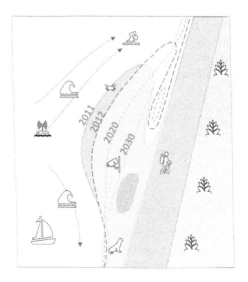

FIGURE 4.10
Sand Engine infographics. This shows spreading and migration of the sand to the north and the south. (Modified from www.dezandmotor.nl/uploads/2015/09/infographic-zandmotor-eng.pdf.)

the hydrodynamics be affected around the area and would any changes be dangerous for swimmers and surfers? To this end, monitoring of currents, waves, sand transport and evolution of the seabed, groundwater changes, and flora and fauna were undertaken from the planning stages in order to assess the impact of the project from the outset. The evolution and interaction of these forcings, and of the coastal habitats, with the Sand Engine, has been closely studied since 2011. After 5 years, the sand in the Sand Engine spread along the coast in a satisfactory manner, and it was concluded that the scheme would work for a longer timescale than the 20 years for which it was originally planned. The wave and winds moved the sand in the north-easterly direction, feeding sand into the coastal dunes backing the system. Storms reshaped the system in different ways, for example, by opening new channels between the lagoon that was built in the artificial peninsula. This lagoon became more elongated and shallower as time passed. The dune system behind the Sand Engine seemed to form much slower than expected, indicating that the effect of the wind on sand transport was actually smaller than that originally estimated. However, this dune system was being reinforced before the Sand Engine project began, and therefore the system was not affected negatively by the slower-than-expected replenishment of sand from the Sand Engine. The Sand Engine was colonised by several sea grasses and protective plant species. As for the marine habitats, after 5 years, it was difficult to determine whether the scheme was better for marine habitat restoration than traditional renourishment schemes.

In terms of attractiveness to visitors, the site has been successfully used by water sport enthusiasts as well as beachgoers who visit the area because the beaches are wide and quiet. Other Sand Engine projects along the Dutch coast, and in Norfolk, UK, are currently being planned. A picture of the Sand Engine seen from the North is shown in Figure 4.11 (from https://beeldbank.rws.nl and made available with permission by Beeldbank Rijkswaterstaat/Joop van Houdt).

An significant number of stakeholders have been involved in the Sand Engine project, and while local authorities have been responsible for organising management and supervision, there has been little involvement of local NGOs after the start of the project. This has been a disadvantage, because their involvement would have helped focus some of the habitat creation objectives much better (Buitenkamp et al. 2016). Moreover, the monitoring activities were partially funded by a monitoring group, and European Regional Development Funding, including the EcoShape project. Coordinating all research activities has proved more challenging than initially estimated, and the 5-year policy report recommended that the research outcomes and the success of the project inform future soft engineering strategies, ensuring that multifunctionality and joint financing are considered seriously from the beginning (Buitenkamp et al. 2016).

From the fieldwork perspective, it is worth noting the excellent review by de Schipper et al. (2016) on beach renourishment activities from the 1900s

FIGURE 4.11
Picture of the Sand Engine from 2012: picture no. 467068. (From
https:// beeldbank.rws.nl, © Beeldbank Rijkswaterstaat/Joop van Houdt—
reproduced with permission.)

onwards, but the core of their study focused on a detailed analysis of the evolution of the Sand Engine from an observational oceanographic and engineering
perspective for the first 18 months of the project. Importantly, de Schipper
et al. (2016) commented that the success of a beach renourishment scheme
depended, principally, on the amount of sediment that remains in the project
area over time. However, in the case of the beach engine, it is not the ability of the sediment to stay in the area which counts, but the ability of the
megarenourishment to spread along the coast and protect the beach within
the municipality. Therefore, the sediment dynamics and shoreline evolution
of the scheme had to be evaluated in detail in order to understand how the
sediment spreading processes, and whether all the beaches which need to be
nourished are benefiting sufficiently from the scheme, as required by the Dutch
government (Van Koningsveld & Mulder 2004, Mulder et al. 2011). Therefore,
a monitoring programme is absolutely crucial to assess the rate of spreading
of the sand and quantify the volumes of sand being deposited at different sections of the coast. A baseline study is also necessary so that the beach and
the forcing conditions are fully characterised prior to the start of the scheme.
The extent of the coast that has to be characterised is limited, in general, to
the coastal cell that will be affected by the scheme. In the case of the Westland
Sand Engine, this corresponds to the Westland coastal cell, between Hoek van
Holland and the jetties of the Scheveningen marina, shown in Figure 4.12, with
the local wave climate rose presented in Figure 1 of de Schipper et al. (2016).

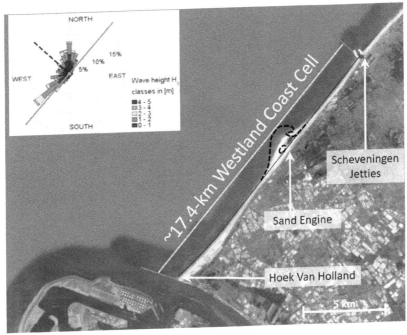

FIGURE 4.12

Location of the Westland Sand Engine and local wave climate rose. (Modified from de Schipper et al. 2016.)

The Sand Engine initially consisted of one meganourishment peninsula with 18.7 M m³ of sediment in the form of a hook (shown as a black dashed line in Figure 4.12), with the curved end facing north, and two shoreface nourishments of 2 M m³ and 0.5 M m³ to the north and south of the peninsula, respectively. However, the term 'Sand Engine' is generally understood to refer to the peninsula alone (de Schipper et al. 2016). The purpose of beach nourishments is twofold, on the one hand, the added sediment acts as a wave dissipation scheme and, on the other, as cross-shore and alongshore sediment feeder. The Sand Engine is principally an alongshore feeder project in the initial phase, when the sand was placed in an area of around 2,500 m alongshore and 1,000 m cross-shore. Because of this design, the peninsula has a very concentrated volume of 10 m³/m alongshore, whereas other projects have cross-shore volume sizes under 2 m³/m alongshore. Also, the peninsula has very sharp angles on the coastline so that sediment is redistributed rapidly by the longshore currents along the Westland coastal cell (de Schipper et al. 2016). The background image in Figure 4.12 shows the current shape of the Sand Engine as it appears from the most recent satellite image of ©Google Earth, which seems to be from 5th of July 2018. Figure 4.13 shows a composite photo illustrating the evolution of the Sand Engine also as seen from space with ©Google Earth, from the same angle and same height above ground

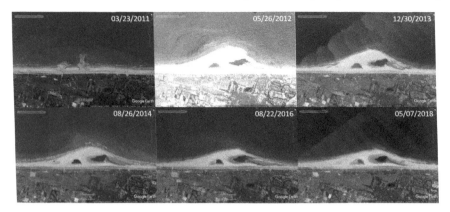

FIGURE 4.13
Evolution of the Westland Sand Engine from March 2011 (beginning of construction) to May 2018. The shoreline around the Sand Engine is rotated by $-50°$ with the horizontal. The images show a spread of the sand predominantly towards the northeast, consistent with a bimodal wave climate with predominant waves coming from the southwestern and northern directions, moving the sand predominantly northward and towards the shore.

(© Google Earth). Although the MSL may not correspond with the shoreline in all pictures, they have been chosen to show the temporal, progressive redistribution of the sediment in the scheme along the coastline due to the action of waves, tides, and wind blowing. The wave conditions were measured with a directional 'Europlatform' buoy located 40 km offshore at 32 m water depth. The wave rose shown in Figure 4.12 indicates that the waves come from two predominant directions. Waves from the southwest are the most dominant, followed by waves coming from the north. This wave climate causes a spread of the Sand Engine's material towards the beach and predominantly in the northeastern direction. This is consistent with the temporal evolution of the hook shown schematically in Figure 4.10 and the evolution of the Sand Engine that is observed from the satellite imagery reproduced in Figure 4.13.

Monthly *in-situ* beach surveys, and biannual LiDAR measurements, showed that at least 20% of the sediment was transported by the wind towards the backshore dunes (de Schipper et al. 2016). The mean tidal range at the site is 1.7 m with a spring–neap variation between 1.48 and 1.98 m. Horizontal tidal velocities are of the order of 0.5 m/s. The wave climate has a seasonal cycle with strongest waves between September and December. The maximum daily-averaged wave height recorded during the recording period was 4.4 m during a storm in January 2012. During the summer, wave heights were low in general with a lowest average value of 0.7 m between two monthly surveys. Sediment size analyses during the construction phase indicated an average median grain diameter d_{50} between 182 and 242 μm with a standard deviation of 50 μm (de Schipper et al. 2016, Luijendijk et al. 2017).

4.3.2   Process-Based Modelling of the Sand Engine with Delft3D

Although several modelling efforts were undertaken to analyse and predict the dynamics and long-term evolution of the Sand Engine, here we will focus on the work by Luijendijk et al. (2017), who used the process-based model to hindcast the evolution of the Sand Engine between August 2011 and August 2012, Delft3D. Delft3D was used in vertically averaged mode to assess whether it could reproduce the evolution of the Sand Engine during this period, and to analyse how each of the different forcing mechanisms might be affecting such evolution. This can be done with numerical models only, as with field monitoring campaigns, one can only analyse the evolution due to all the forcings combined, sometimes in ways which are difficult to interpret. Also, the numerical model helps the understanding of the evolution of the human-designed peninsula into a sandy headland that has been, and continues to be, shaped by the environmental forcings. A schematic of the model conceptual framework and implementation is shown in Figure 4.14. The Delft3D model implemented is a nested model with three domains. The highest-resolution domain (HRD) is centred in the Sand Engine and is 9,500 m long (in the longshore direction) and 4,000 m wide (in the cross-shore direction). The cell size varies from 35 to 135 m; the finest cells cover the Sand Engine, and then they increase in size further from it. The intermediate resolution domain (IRD) covers a coastal strip, and the large-scale domain is a continental shelf domain (CSD).

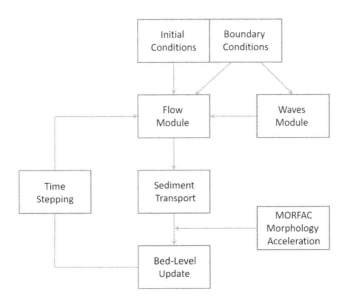

FIGURE 4.14
Schematic of Delft3D model implemented by Luijendijk et al. (2017).

The CSD was forced at the seaward (outer) boundary with thirteen principal components of the astronomical tide. The storm surge levels observed at the tide gauge located at Hoek Van Holland were added to the tidal forcing. At the lateral open boundaries, zero-gradient alongshore water-level conditions were applied (Roelvink & Walstra 2004). The wind speed forcing conditions applied on the model domain were those from the meteorological station located in Hoek Van Holland as well, which assumes that the wind speed and direction changes little within the domain. The CSD is 79,000 m long and 42,000 m wide (cross-shore direction). Its resolution varied from 170 to 2,000 m. The IRD, on the other hand, is 25,000 m long and 13,000 m wide. The HRD is divided into a hydrodynamic domain and a wave domain. The wave domain is similar to the hydrodynamic domain but is extended by 500 m on all three open boundaries (to the south, the west, and the north). All three models were run in 2D and in hydrostatic mode.

In order to validate the model, a 3-month simulation with default parameter values was run, and the tidal and nontidal depth-averaged water levels and currents were separated using the t_tide software developed by Pawlowicz et al. (2002). The first month is used as spin-up time of the simulation. The first six dominant tidal principal components predicted by the model were compared with those obtained from ADCP (acoustic Doppler current profiler) data at a location close to the Sand Engine, with good agreement between model and measurements—the authors mention a discrepancy of <0.02 m in amplitude and <12 min in phase for all six components for the water levels, which corresponds to a root mean square error (RMSE) of 14%. The components corresponding to the tidal currents were also compared, and for the M_2 constituent, they found an RMSE of 3.3%. As for the waves, there was also good agreement between modelled and observed wave heights for the period of 5 weeks for which the two datasets were compared. The agreement was not so good in mild wave conditions, but as the authors highlight, these mild conditions are less relevant for morphological evolution studies than stormy periods would be. In fact, the authors report that after 1 year of simulation, they observed that the bed-level changes were significant only if the wave heights were larger than 1.5 m.

Once the validations were performed and the performance of the model deemed satisfactory, the full-year simulation was run with morphological factor (MORFAC) set to 1, corresponding to a simulation with no morphological acceleration—this is known as a *brute-force* simulation. As a result of these validations and the brute-force simulation, all wave conditions with waves heights below 1.5 m and wave directions away from the Sand Engine were removed from the computation, corresponding to 40% savings in computational time. These waves had very little effect on the evolution of the Sand Engine, and this is the reason why they were removed.

As well as the default settings, more advanced options of the Delft3D model were tested, including the type of wave module (SWAN or wave roller model), the values of the bed roughness (either fixed or temporal and space varying), the horizontal eddy viscosity (ν_H), the horizontal eddy diffusivity (D_H), the wave breaking parameter (γ), the median grain size sediment transport formulation (d_{50}), and the factor of erosion for adjacent cells ($ThetSD$). The advanced options selected improved the model predictions significantly but were more computationally expensive, mostly due to the selection of optimal transport formulation (Van Rijn et al. 2004), which is based on an iterative process. A full-year simulation took 20 days to run on a desktop PC with a 3.40 GHz core. $ThetSD$, the roller model, and the transport formulation were the three factors that most affected the model results.

The model hindcast for the period August 2011–August 2012 was created in order to have a model that can reproduce accurately the magnitude of the volume changes at the stretch of coast where the Sand Engine is located and the erosion and accretion patterns observed along the entire study area over that period of time. The volume changes were computed in three control sections: the southern section (S Sec), the peninsula, and the northern section (N Sec), which are shown in Figure 4.15, together with the wave rider buoy and the two nearshore ADCPs used for the model validation that was described above. The volume changes in the three sections, the sum of the southern

FIGURE 4.15
Control sections defined by Luijendijk et al. (2017) with nearshore instrumentation.

and northern sections (N+S), and the volume change for all three sections are shown in Figure 4.16a. The Brier Skill Score (BSS), defined as (Sutherland et al. 2004)

$$\text{BSS} = 1 - \frac{\text{MSE}}{\text{MSE}_{\text{ini}}}, \tag{4.13}$$

is used to assess the performance of the model. MSE is the mean-squared error, and MSE_{ini} is the MSE of the zero change reference model, that is, the reference prediction for which the initial bed is taken. The *BSS* as defined is the model added skill relative to a prediction that nothing changes. A prediction that is as good as the zero change reference prediction ($\text{MSE} = \text{MSE}_{\text{ini}}$) receives a score of 0, and a perfect prediction ($\text{MSE} = 0$) a score of 1. A BSS over 0.5 is considered to give an excellent measure of performance of the model in relation to the observations. As can be seen from Figure 4.16b, this *BSS* is obtained from month no. 8 onwards, corresponding to April 2012. Between December 2011 and March 2012, the performance of the model is good, with BSS values between 0.2 and 0.5. These results show that the Delft3D model developed performs well in relation to the volume changes that the model predicts, at least from December 2011 onwards. That the BSS increases over time is expected; it means that the model reproduces well the erosion patterns due to the natural evolution of the Sand Engine away from the initial state and the resulting increase of MSE_{ini}.

In relation to the erosion and accretion patterns observed in the field and those predicted by the model, it was also found that the model performed well for this criterion, in particular for the following features of the Sand Engine:

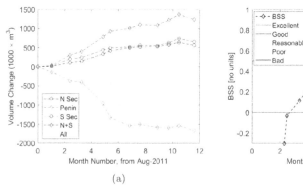
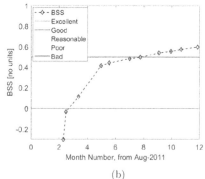

(a)   (b)

FIGURE 4.16
Volume changes and BSS values from August 2011 to August 2012. N Sec: northern section, Penin: peninsula, S Sec: southern section, N+S: sum of adjacent sections, All: sum of all sections. (a) Observed and predicted volume changes and (b) BSS and its ranges. (Modified from Luijendijk et al. 2017.)

- the development and evolution of a spit and a channel on the northern side of the Peninsula

- the pattern of erosion of the most seaward part of the Peninsula

- the extent of the erosional area and a shoreline retreat of about 100 m

- the transition zone between erosion and sedimentation in the northern part

- the sedimentation pattern in the southern part

Other features are smoothed out in the model, such as some of the submerged bars at NAP = −3 m or the cross-shore dominated bar behaviour in the northern section. However, this is unavoidable because of the 2D nature of the model. Finally, the modelled plan shape evolution was sensitive to the numerical scheme to drive the erosion of the dry areas and the applied sediment transport formulation. However, the model and data intercomparison showed that the model reproduces well the sediment redistribution patterns.

Another meganourishment is discussed in Tonnon et al. (2018), which complements the work of de Schipper et al. (2016) and Luijendijk et al. (2017). Tonnon et al. (2018) analysed both the Sand Engine, and a project near Petten, the Netherlands, with a renourishment volume of 40 M m^3, that is, twice the volume size of the Sand Engine. Petten is around 100 km north of the Sand Engine scheme but is not a spit of sand that spreads north and south of the coast throughout several years as the Sand Engine. In fact, the Petten project is part of the primary coastal defence in that area and requires to be renourished periodically to maintain its size, shape, and function. In the Petten beach nourishment case, numerical morphodynamics modelling is still useful to analyse and plan better the beach nourishment requirements of the scheme. The reader is referred to Tonnon et al. (2018) for more information.

4.4 Offshore Renewable Energy

The fundamental numerical modelling and physical processes associated with the evaluation of marine and offshore renewable energy resources are the same as for other applied science and engineering projects. Monthly, seasonal and annual cycles are of importance, as in other research fields, but the analysis of periodic patterns within the signal is further complemented with studies of power densities and their statistical distributions and estimates of annual energy production. Power histograms or power matrices reveal crucial information regarding the frequencies of occurrence of a certain power at given environmental conditions, as well as low- and high-power thresholds. These analyses inform economic viability and survivability conditions at potential development sites. These power data visualisations help identify the suitability of a test or development site, as well as the best technological choices. Therefore, they are extremely important to developers and decision makers. Here we are interested only on research focused on environmental impacts and, in particular, on sediment transport and regional morphodynamics. So we will make an approximate assessment of the number of papers published on these topics since 1980 and, in some cases, present a sample study. This section focuses on offshore technologies that transform kinetic energy into mechanical energy and that which, in general, can be installed close to the coast, namely, in-stream or hydrokinetic energy, wave energy, and offshore wind.

4.4.1   Hydrokinetic Energy

In-stream technologies exploit the kinetic energy in moving water and turn it into power. The devices feed electrical energy to the grid in a relatively simple way. These technologies can be deployed in rivers, artificial canals, estuaries, or nearshore coastal areas. However, the technical specifications usually are very restrictive in relation to the velocity ranges and minimum water depths or channel widths. For example, for some technologies deployed in rivers and canals, an optimal velocity range of 1.5–3.5 m/s, a minimum water depth of 2 m, and minimum channel width of 4 m (Instream Energy Systems 2018) are specified. In coastal areas, maximum water depths are also important. However, this is a small additional consideration for coastal and ocean currents, because estuarine and coastal ocean currents behave very differently to river currents.

Tidal turbine arrays are known to affect the flow patterns around them by reducing the tidal speeds within the array and increasing them around it and by changing the direction of the tides in and near the region where they are located (Ahmadian et al. 2012). These changes also have impacts on the SS or contaminant concentrations, which have been shown to be slightly lower within the tidal array. Both the SS and contaminant concentrations have been found to be lower both upstream and downstream and higher at the sides of the array, with the impact extending to up to 15 km away (Ahmadian et al. 2012). However, the effects on water level are generally small, in the order of millimeters or <10 cm in water depths of the order of 20 m, as found by Ahmadian et al. (2012) for the Severn estuary (study site shown in Figure 4.17) and by Defne et al. (2011) for the Canoochee and the Ogeechee rivers (study site shown in Figure 4.18), respectively; such changes in water levels are negligible. On the other hand, some studies suggest that the energy extraction should be limited to 15% of the total energy within the system in order to avoid significant impacts; however, other studies suggest that up to 30% of the energy could be extracted without important consequences. In order to assess the hydroenvironmental impacts of power extraction, Defne et al. (2011) analysed the effects of an array of tidal energy converters on estuarine hydrodynamics, using regional models and applying them at the

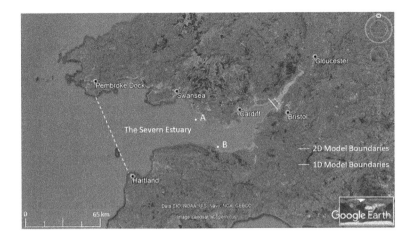

FIGURE 4.17
Severn estuary model domain. (Modified from Ahmadian et al. 2012.)

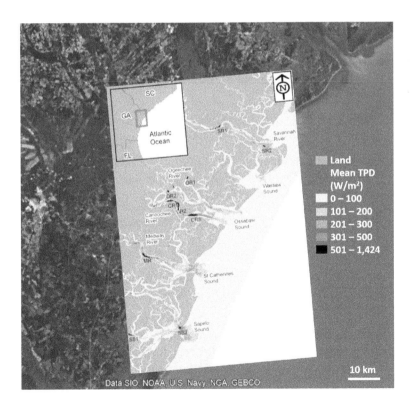

FIGURE 4.18
Georgia State (USA) Study Site. (Modified from Defne et al. 2011.)

Canoochee and the Ogeechee rivers, in the state of Georgia, Western USA. For the Canoochee river, the authors considered an array that extracted energy at four model cells across the rivers; the depth at these cells ranged between 3.5 and 8.4 m, and the authors kept the maximal kinetic energy extraction to 20% (case 1) of the total kinetic energy in the undisturbed river flow. Then the energy extraction is increased to 45% (case 2), and the effects on the far-field hydrodynamics are analysed. The authors find that the current magnitude is practically not affected in case 1 and negligibly small in case 2. Power density is observed to be reduced during periods with high flow speeds, because of the power density dependence on the cube of the speed. The authors were surprised to find that the sum of the residual power and the extracted power during extraction was larger than the energy available in the baseline (undisturbed) case. This indicated that there is a transfer of potential energy to kinetic energy when the tidal turbines are introduced into the system and hence a redistribution of power. Other effects of the tidal turbines on the hydroenvironmental characteristics include the following:

- The mean tidal stream power density is affected near the region of extraction, particularly downstream of the power extraction area in the direction of ebb currents.

- In the Canoochee river, the tidal currents are stronger during ebb tides than during flood tides, and therefore, more power is extracted during ebb tides.

- There is a power density extraction asymmetry around the line of extraction, with larger effects observed seawards.

- The impacts are observed up to 5 km away from the power extraction location, where the change in power extraction for case 1 in relation to the baseline case is of 10%.

- In case 1, there is a mean power extraction increase of about 30 W/m^2 in the northern edge of the extraction location, but in case 2, there is also an increase in this edge but of about 62 W/m^2.

- In case 1, the tidal current magnitude upstream of the power extraction location decreases by 0.07 m/s in the flood direction during flood tides.

- Stronger currents during ebb tides allow for more power extraction and a larger current magnitude drop during this part of the cycle. The maximum drop is of 0.45 m/s, but this diminishes within a few kilometres.

- Overall, it can be said that there are no major changes in tidal current speeds between case 1 and the baseline, except at the location of the tidal power extraction.

- Some impacts on water-level maxima, minima, and phases are observed, but the effects are extremely small (of the order of centimetres), in particular for case 1. The effects for case 2 may be acceptable if the impact on the wetland areas is small, and thus the ecological implications are small too.

In terms of the modelling approaches for tidal extraction and the hydroenvironmental impacts of tidal arrays, it is worth noting that some models include retarding or drag forces in the momentum equations, such as the models of Ahmadian et al. (2012) and Defne et al. (2011) discussed above, and other models implement tidal devices within shallow-water models using the actuator disc concept (Magar 2018), which is equivalent to adding a quadratic momentum loss term in the momentum conservation equations. The actuator disc concept is equivalent to adding a volumetric sink term in the momentum conservation equations. Here the thrust coefficient is linked to the equivalent coefficient in the Linear Momentum Actuator Disc Theory; the thrust coefficient represents the level of extraction of energy from a tidal energy converter. In a recent publication, Chatzirodou et al. (2019) implemented the actuator disc approach in a 3D Delft3D model at the Pentland Firth Inner Sound Channel, in Northern Scotland (study site shown in Figure 4.19), to analyse the effect of a tidal array on the morphodynamics of the area and, in particular, on the three sandbanks located in the vicinity of a proposed tidal array development site. In previous papers, these authors found that the sandbanks are very active, with cumulative bed-level changes that can be of up to 2 m in a spring–neap tidal period (Chatzirodou et al. 2016).

Chatzirodou et al. (2019) designed a multiscale model grid that varied in cell size from 2 km in the outer boundaries, to 200 m in the deep regions of the Pentland Firth, to 20 m in the shallow areas near the coast of Stroma,

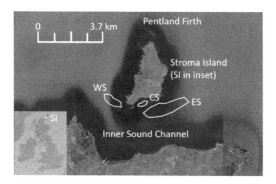

FIGURE 4.19

Pentland Firth Inner Sound Channel study site. WS is the western sandbank, CS the central sandbank, and ES the eastern sandbank referred to in the text. (Modified from Chatzirodou et al. 2019.)

where one of the sandbanks is located. The model has one layer in the coarse-resolution domain and ten vertical layers in the medium- and fine-resolution domains. At the open boundaries, the coarse domain is forced with eight tidal principal components, four diurnal and four semidiurnal. The sediment transport formula applied is the bedload transport formula of Van Rijn (1993), and it is applied in the region of the sandbanks, as the bed is nonerodible elsewhere. Five scenarios, with thrust coefficients varying from 0.18 to 0.85, were tested. The value of 0.85 is the thrust coefficient value that is commonly assumed for a generic tidal energy converter with a diameter of 20 m. The results show that, in some areas, the surface velocities may be up to four times higher with the tidal array in place and that, in others, the surface velocities may decrease by 25%. The velocity magnitudes change a lot more at the surface than at the bottom. Bottom velocities increase by a factor of three in the northwest part of the array and decrease by 23% in the westward wake region relative to the northeast currents. The water-level change is <20 cm, which can be neglected as the average local water depths in the regions vary between 25 and 75 m.

The residual tidal currents, on the other hand, may be reduced by up to 1 m/s in some regions, and this can have a significant effect on the sediment transport, which generally is driven by the residual flows. The results obtained were based on the worst-case scenario of Chatzirodou et al. (2019), with a thrust coefficient of 0.85. Patterns of erosion and deposition were changes by the effects of the tidal array on the residual flow and residual sediment transport patterns. There was a significant bed-level change of the sandbank on the eastern side of the Pentland Firth Inner Channel, of up to 1 m after a 1-month simulation. In the sandbank on the shallow central region, just south of Stroma Island, the residual transport rates decrease by more than 50% over the whole central sandbank, resulting in an increase in sandbank bed level of up to 0.7 m. Finally the third sandbank, on the western side of the Channel, is on the lee side of the Stroma Island, where there is very little flow separation, and as a consequence, the tidal array has very small impact on the residual transport rates in this region. The sandbank accreted by a mere 0.14 m over the whole sandbank area, relative to the baseline. These bed-level changes, particularly on the eastern and central regions of the Pentland Firth, could affect the long-term performance of the tidal array and may even require sand dredging campaigns. However, it is important to recall that this is the worst-case scenario.

4.4.2 Wave Energy

The wave energy resource is characterised by the wave power density, P_w, which is determined as the integral, over all wave frequencies and directions, of the product of the wave energy group speed and the energy spectrum of the waves, $E \propto H_s^2 T_e$, where H_s is the significant wave height and T_e is the wave energy period. These wave spectrum variables can be expressed in terms of the frequency, f; the distribution of the energy in different directions, θ; and the wave energy spectrum directional coordinate. The overall wave power is

$$\iint E(f,\theta)C_g df\, d\theta,$$

where C_g is the group velocity of the waves and

$$E(f,\theta) = \frac{1}{8}\rho g H_s^2 T_e,$$

as defined in Reeve et al. (2011). Therefore, three parameters and their distributions affect the wave power density, and so joint probability analyses have to be performed in order to characterise the wave energy resource adequately. In relation to wave energy computations, several studies focus on assessing the most energetic field conditions based on the distribution of wave heights only, but joint H_s–T_e diagrams are significantly more meaningful. Although storms generate the largest waves, such waves are also in many cases dangerous for the survivability of the devices. The H_s–T_e diagram can also help assess how frequently the wave energy conditions are above the threshold of device survivability. An H_s–T_e diagram, obtained for the Wave Hub site (shown in Figure 4.20), using the wave conditions generated by a WAVEWATCH III (WW3) model for the Bristol Channel at a resolution of $0.2° \times 0.2°$, is shown in Figure 4.21. The H_s–T_e diagram

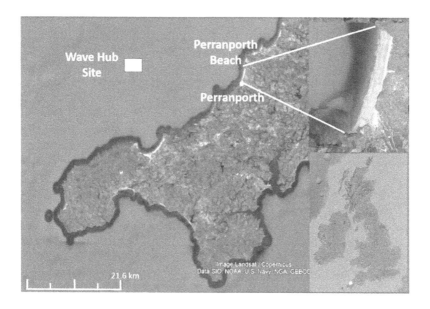

FIGURE 4.20
Wave Hub site and Perranporth beach. (Based on Abanades et al. 2014.)

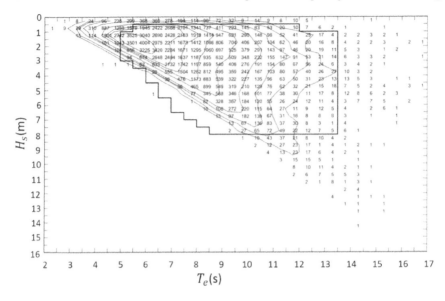

FIGURE 4.21

H_s–T_e diagram for Wave Hub site using wave conditions from a 40-year simulation with WW3. (Reproduced with permission from Reeve et al. 2011.)

shows that, at the Wave Hub site, the most common waves have periods between 5 and 8 s and wave heights between 1 and 3 m. The box corresponds to the Pelamis Power Matrix and shows that most wave conditions at the Wave Hub are within the operational H_s–T_e range for Pelamis, but some of the waves have higher wave heights, or larger wave periods, than those in the Power Matrix.

Now, because wave farms extract energy from the environment, a milder wave climate is observed on the lee side of wave farms (Millar et al. 2007). This affects the hydrodynamics and morphodynamics of nearby coastal regions and may serve as a coastal erosion mitigation measure, because the incoming waves have less energy to dissipate and contribute to sediment transport (Iglesias et al. 2018). In this way, wave energy conversion would help secure and diversify the energy supply, on the one hand, and reduce beach erosion problems in some coastal environments, on the other. In the study by Abanades et al. (2014), the authors assessed the potential effects of a wave farm on the beach morphodynamic processes at Perranporth, UK (shown in Figure 4.20), a beach close to the Wave Hub site. Abanades et al. (2014) coupled a high-resolution spectral wave model with the morphodynamic model XBeach in a 2D configuration, so the evolution of the full beach could be investigated under storm conditions and in the presence of a wave farm. They found that, under storm conditions, the wave farm could reduce beach erosion by up to 50%, demonstrating that wave farms can act as coastal defences. However, considering the large costs associated with wave energy converter (WEC) development, trialling, and commercialisation, as a WEC developer, would you put your assets at risk?

4.4.3 Offshore Wind

According to the Web of Science, more than 12,312 papers on offshore wind have been written since 1980. However, there have been step increases in the number of papers published per year, with less than ten papers per year, from 1980 to 1990; 69–140 papers per year, between 1991 and 2000; 162–426 papers per year, between 2001 and 2010; 567–1297 papers per year, between 2011 and 2018; and 447 papers written in the first half of 2019 (Web of Science 2019). However, within this broad research topic, only 476 papers have focused on environmental impacts, with three universities accumulating 50 research papers: the Norwegian University of Science and Technology (NTNU) (24), the University of Plymouth (14), and Delft University of Technology (12). Most of these outputs are the result of a number of projects that have been approved by either National or EU research councils. For example, during her time at the University of Plymouth, and later at the Centre for Scientific Research and Higher Education of Ensenada (CICESE), the author participated in the INTERREG project 'Offshore Foundations' Environmental Impact Assessment' (OFELIA)—see www.keep.eu/project/7983/ofelia for more details. This project focused on hydrodynamic and sediment transport impacts of offshore wind turbine monopile foundations. In particular, several physical modelling studies at laboratory scales were carried out in all three partner universities (the University of Caen, University of Le Havre, and University of Plymouth). A numerical modelling exercise at regional scales was also performed using the Courseulles-sur-Mer wind farm site as case study (Rivier et al. 2016).

The regional hydrodynamic model used in OFELIA was the MARS3D model of the 'Institut Français de Recherche pour l'Exploitation de la Mer' (IFREMER) in France, developed by Lazure & Dumas (2008). MARS3D is applicable to coastal bays and estuaries, with a coupling between barotropic and baroclinic modes which allows for consistent sediment and pollutant transport between the 2D and the 3D configurations of the model. Like Delft3D, MARS3D is by default a terrain-following model with vertical layers corresponding to normalised isopycnal layers of constant density, as introduced by Phillips (1957) and Bleck & Smith (1990). The depth normalisation helps maintain high resolution in the model in a wide range of water depths. Some shallow-water models use a time splitting scheme that solves implicitly the internal waves (the internal mode) with large timesteps and explicitly the free surface (gravity) waves (the external mode) with small timesteps. The two modes have a two-way coupling: the barotropic pressure gradient is prescribed by the external to the internal mode, and bottom stress, integrated pressure gradient, and internal stresses are provided by the internal to the external mode (Lazure & Dumas 2008). However, other implicit or semi-implicit models can be implemented to solve the equations for the external, barotropic mode; in this case, the internal and external modes can be solved with large timesteps, as the CFL stability conditions are removed. MARS3D uses an Alternative Direction Implicit (ADI) scheme (Lazure & Dumas 2008) and thus falls into this latter category.

The sediment transport module in MARS3D is based on the advection–diffusion equation formulated by Hir et al. (2011), with a bottom boundary condition involving erosion and deposition processes which are adequate for sand and mud mixtures. In Rivier et al. (2016), the sediment fluxes, deposition and erosion rates, and settling velocity were determined using a current flume experiment set-up, where the sediment grain sizes varied between 140 and 450 μm. They considered only the SSC with a fixed bed level in the hydrodynamic module but a variable spatio-temporal availability of sediments inducing a change of bed thickness in the sediment transport module. The original contribution of Rivier et al. (2016) consists on the implementation in MARS3D of a drag force, $\mathbf{F}_D$, per unit area induced by the monopile, defined as

$$\mathbf{F}_D = -\frac{1}{2}\frac{\rho_w C_D D}{\Delta x \Delta y} \parallel \mathbf{u}_\infty \parallel \mathbf{u}_\infty, \qquad (4.14)$$

where ρ_w is the water density; C_D is the drag coefficient; D is the monopile diameter; Δx and Δy the cell sizes in the x and y directions, respectively; $\parallel \cdot \parallel$ is the L^2 norm; and $\mathbf{u}_\infty$ (u_∞, v_∞) is the undisturbed current velocity upstream of the pile. The drag force terms are added to the conservation and the turbulence closure equations of the model. The effects of the monopile on the hydrodynamics and the sediment transport are evaluated using the relative difference,

$$\Delta V = \frac{V_{\text{mod}} - V_{\text{ref}}}{V_{\text{ref}}}, \qquad (4.15)$$

for each variable V of interest, with V_{mod} the values with the monopile, and V_{ref} the reference (undisturbed flow) values. The impact of the monopiles was evaluated using the new implementation and the default, dry cell approach already available within MARS3D. Two monopile diameters of 6 and 15 m were also analysed. In the first case, the monopile covers four cells and appears as a square in an x–y plane, whereas in the second one, the monopile covers 21 cells. The analyses were based on four monopiles forming a square, with a side of 950 m.

The main conclusions of Rivier et al. (2016) were the following:

- Currents accelerate on the upstream side of the monopile and decelerate on the downstream side.

- In the regions with larger current speed and shear stress, erosion and sediment resuspension are observed.

- Sediment deposition occurs in regions with weak bottom shear stress as expected.

- If a monopile wake reaches another monopile, then it will affect the hydrodynamics and sediment transport in the vicinity of that monopile too.

The conclusions above are obtained with both the code with drag forces and the code with dry cells at the positions of the monopiles. The main difference between the two methods is that erosion and deposition are higher close to the monopile with the parameterisation method than with the dry points method. For both approaches, the monopile wake can be observed even 2 km downstream of the monopile, and thus it will affect other monopiles in the array if they are in the wave path.

5

Epilogue

Our journey has come to an end. In these pages, we attempted to provide comprehensive coverage of sediment transport and morphodynamics modelling for coasts and shallow environments, with an emphasis on the physical processes, the modelling design, and the more significant results, at different dynamic scales.

We started describing the dynamics of the coastal environment from the microscopic to the macroscopic scales, possibly because it was more intuitive to do it like so. We also strived to describe in detail the physical processes first and then see how the physics are implemented in numerical models.

In the chapter on fundamental physics (Chapter 2), we start with sediment composition concepts and move on to sediment transport formulations, with descriptions of the flow characteristics in the bottom boundary layer. We then analyse the processes responsible for sediment transport, both in the deep ocean and in coastal waters, with a focus on the effects of waves and tides on sediment transport and morphodynamics. We evaluate in detail the roles that waves and tides play in the evolution of open coasts and sheltered coastal ecosystems. We also assess the differences in sediment composition and vegetation type and coverage between these two contrasting coastal environments.

In the chapter on numerical modelling (Chapter 3), we place significant emphasis on good modelling practices, particularly for shallow-water regional models. However, these practices apply to other model types, such as Reynolds-Averaged Navier–Stokes (RANS), Large-Eddy Simulation (LES), or Direct Numerical Simulation (DNS) models. We start with mesh development issues, such as a presentation on structured and unstructured grids, suggestions on grid resolution needs according to the study objectives, and properties of sigma-level and z-level meshes. We then comment on different bathymetries of global coverage available online, as well as high-resolution bathymetries obtained with *in-situ* and remote sensing measurements. A discussion on tidal, wave, and meteorological forcings follows, together with a description of different models that couple hydrodynamics, sediment transport, and morphodynamics. We close this chapter with some case studies on numerical modelling for complex environments. The final chapter (Chapter 4) contained a series of examples that illustrate novel and future applications of morphodynamics modelling. These examples focus principally on

- the evaluation of potential climate change impacts on the coast and climate change mitigation strategies,

- new coastal mega beach nourishment projects that reduce the need for regular interventions and thus reduce the costs of coastal protection, and

- studies on potential environmental impacts of offshore renewable energy farms.

We would have liked to have been able to expand more on several modelling approaches that unfortunately we covered only superficially. Such is the case, for instance, for Boussinesq-type models. The effects of meteorological forcings on the evolution of complex environments should also expand in future editions. We engaged, mostly, on the description of regional models forced by tides and by spectral wave formulations, not because they are the best models but because they are used extensively in industry and they provide useful insights into the short to medium-term evolution of coastal environments. Another possible weakness remains in the presentation of the most significant results concerning previous research. However, I hope to have corrected this in most cases.

Climate change will have predicted and unforeseen impacts on coastal environments. Therefore, as coastal modellers, it is our responsibility to envision different scenarios and assess the effects of extreme events on coastal ecosystems and urbanised coasts alike. While the economic costs of damage to coastal cities are evident, coastal ecosystems provide vital services to humankind. Offshore reefs, for example, protect the coast from erosion by incoming storms through dissipation of wave energy. Mangroves and salt marshes also dissipate wave energy and, at the same time, act as sediment traps. Sea-level rise will force us more and more to design managed realignment schemes to reduce the costs associated with the protection of our assets.

In the 20th century, humankind tended to use hard engineering structures for coastal protection, which led to the loss of whole coastal towns, due to a lack of detailed understanding of the sediment pathways. However, we seem to have realised that working with nature will provide better long-term solutions to coastal erosion and flooding. Twenty-first century coastal protection designs are a reflection of this new philosophy. This is evident, for example, in the case of the floating gardens and play areas designed to protect the coastal sectors of New York City.

Finally, we concluded with environmental impacts of offshore structures, in particular offshore renewable energy devices, and with a focus on devices that extract kinetic energy such as wave energy converters, wind energy farms, and in-stream tidal energy converters. We analysed some of their effects on nearby bedforms and coastlines, which are still poorly understood. A further topic that is still missing here is the greenhouse gas storage capacity of salt marshes, mangroves, and other coastal environments. We will attempt to include examples of human interventions that have, or could, promote this storage capacity through managed realignment schemes, in a subsequent edition of this work.

Bibliography

Abanades, J., Greaves, D. & Iglesias, G. (2014), 'Coastal defence through wave farms', *Coastal Engineering* **91**, 299–307. doi: 10.1016/j.coastaleng. 2014.06.009.

Achete, F., Van der Wegen, M. V., Roelvink, J. A. & Jaffe, B. E. (2017), 'How can climate change and engineered water conveyance affect sediment dynamics in the San Francisco Bay-Delta system?' *Climatic Change* **142**(3–4), 375–389. doi: 10.1007/s10584-017-1954-8.

Ahmadian, R., Falconer, R. & Bockelmann-Evans, B. (2012), 'Far-field modelling of the hydro-environmental impact of tidal stream turbines', *Renewable Energy* **38**(1), 107–116. doi: 10.1016/j.renene.2011.07.005.

Alexander, K. S., Ryan, A. & Measham, T. G. (2012), 'Managed retreat of coastal communities: Understanding responses to projected sea level rise', *Journal of Environmental Planning and Management* **55**(4), 409–433. doi: 10.1080/09640568.2011.604193.

Allen, J. (1997), 'Simulation models of salt-marsh morphodynamics: Some implications for high-intertidal sediment couplets related to sea-level change', *Sedimentary Geology* **113**(3), 211–223. www.sciencedirect.com/science/article/pii/S0037073897001012.

Alvarez, F. & Pan, S. (2016), 'Predicting coastal morphological changes with empirical orthogonal function method', *Water Science and Engineering* **9**(1), 14–20. doi: 10.1016/j.wse.2015.10.003.

Álvarez, L. G., Suárez-Vidal, F., Mendoza-Borunda, R. & González-Escobar, M. (2009), 'Bathymetry and active geological structures in the Upper Gulf of California', *Boletín de la Sociedad Geológica Mexicana* **61**(1), 129–141.

Amsden, A. & Hirt, C. (1973), 'A simple scheme for generating general curvilinear grids', *Journal of Computational Physics* **11**(3), 348–359. www.sciencedirect.com/science/article/pii/0021999173900788.

Andreotti, B. (2004), 'A two-species model of aeolian sand transport', *Journal of Fluid Mechanics* **510**, 47–70.

Andrews, D. G. & Mcintyre, M. E. (1978), 'An exact theory of nonlinear waves on a Lagrangian-mean flow', *Journal of Fluid Mechanics* **89**(4), 609–646. doi: 10.1017/s0022112078002773.

António, S. D. (2017), Shoreline and sandbar coupling on a natural and nourished beach, Master's thesis, Department of Physical Geography, Faculty of Geosciences, Utrecht University.

Aranuvachapun, S. & Johnson, J. A. (1978), 'Beach profiles at Gorleston and Great Yarmouth', *Coastal Engineering* **2**, 201–213. doi: 10.1016/0378-3839(78)90020-0.

Atkinson, A. L., Baldock, T. E., Birrien, F., Callaghan, D. P., Nielsen, P., Beuzen, T., Turner, I. L., Blenkinsopp, C. E. & Ranasinghe, R. (2018), 'Laboratory investigation of the bruun rule and beach response to sea level rise', *Coastal Engineering* **136**, 183–202. doi: 10.1016/j.coastaleng.2018.03.003.

Bagnold, R. A. (1966), An approach to the sediment transport problem from general physics, Technical Report 422-1, U.S. Geological Survey. Available at: https://pubs.usgs.gov/pp/0422i/report.pdf.

Balaji, R., Kumar, S. S. & Misra, A. (2017), 'Understanding the effects of seawall construction using a combination of analytical modelling and remote sensing techniques: Case study of Fansa, Gujarat, India', *The International Journal of Ocean and Climate Systems* **8**(3), 153–160. doi: 10.1177/1759313117712180.

Barnard, P. L., Schoellhamer, D. H., Jaffe, B. E. & McKee, L. J. (2013), 'Sediment transport in the San Francisco Bay Coastal System: An overview', *Marine Geology* **345**, 3–17. doi: 10.1016/j.margeo.2013.04.005.

Barnard, P. L., Short, A. D., Harley, M. D., Splinter, K. D., Vitousek, S., Turner, I. L., Allan, J., Banno, M., Bryan, K. R., Doria, A., Hansen, J. E., Kato, S., Kuriyama, Y., Randall-Goodwin, E., Ruggiero, P., Walker, I. J. & Heathfield, D. K. (2015), 'Coastal vulnerability across the Pacific dominated by El Niño/Southern Oscillation', *Nature Geoscience* **8**, 801–807.

Bartholdy, J., Ernstsen, V., Flemming, B., Winter, C. & Bartholomä, A. (2008), On the development of a bedform migration model. In: D. Parsons, T. Garlan & J. Best, (eds). *'Proceedings of Marine and River Dune Dynamics III'*, University of Leeds, Leeds, pp. 9–16.

Battaglia, L., Cruchaga, M., Storti, M., D'Elía, J., Aedo, J. N. & Reinoso, R. (2018), 'Numerical modelling of 3d sloshing experiments in rectangular tanks', *Applied Mathematical Modelling* **59**, 357–378. doi: 10.1016/j.apm.2018.01.033.

Battjes, J. A. (1974), Computation of set-up, longshore currents, run-up and overtopping due to wind-generated waves, Ph.D. thesis, Department of Civil Engineering, Delft University, Delft, the Netherlands, 241 pp.

Benassai, G. (2006), *Introduction to Coastal Dynamics and Shoreline Protection*, WIT Press, Southampton.

Benoit, M., Marcos, F. & Becq, F. (2001), 'Development of a third generation shallow-water wave model with unstructured spatial meshing', *Coastal Engineering Proceedings* **1**(25). https://icce-ojs-tamu.tdl.org/icce/index.php/icce/article/view/5241.

Berberović, E., van Hinsberg, N. P., Jakirlić, S., Roisman, I. V. & Tropea, C. (2009), 'Drop impact onto a liquid layer of finite thickness: Dynamics of the cavity evolution', *Physical Review E* **79**(3). doi: 10.1103/physreve.79.036306.

Bever, A. J. & MacWilliams, M. L. (2013), 'Simulating sediment transport processes in San Pablo Bay using coupled hydrodynamic, wave, and sediment transport models', *Marine Geology* **345**, 235–253. doi: 10.1016/j.margeo.2013.06.012.

Bird, E. & Lewis, N. (2015), *Beach Renourishment*, Springer International Publishing, Cham. doi: 10.1007/978-3-319-09728-2.

Bleck, R. & Smith, L. T. (1990), 'A wind-driven isopycnic coordinate model of the north and equatorial Atlantic Ocean: 1. Model development and supporting experiments', *Journal of Geophysical Research* **95**(C3), 3273. doi: 10.1029/jc095ic03p03273.

Booij, N. & Holthuijsen, L. H. (1987), 'Propagation of ocean waves in discrete spectral wave models', *Journal of Computational Physics* **68**(2), 307–326. www.sciencedirect.com/science/article/pii/002199918790060X.

Booij, N., Ris, R. C. & Holthuijsen, L. H. (1999), 'A third-generation wave model for coastal regions: 1. Model description and validation', *Journal of Geophysical Research: Oceans* **104**(C4), 7649–7666. doi: 10.1029/98jc02622.

Boussinesq, J. (1877), Essai sur la theorie des eaux courantes. In: 'Memoires presentes par divers savants de l' Academie des Sciences de l'Institut de France', Vol. XXIII, Academie des Sciences de l'Institut National, Imprimerie Nationale, France, pp. 1–680.

Boutin, J. & Martin, N. (2006), 'ARGO upper salinity measurements: Perspectives for l-band radiometers calibration and retrieved sea surface salinity validation', *IEEE Geoscience and Remote Sensing Letters* **3**(2), 202–206. doi: 10.1109/lgrs.2005.861930.

Bradford, S. F. (2000), 'Numerical simulation of surf zone dynamics', *Journal of Waterway, Port, Coastal, and Ocean Engineering* **126**(1), 1–13. doi: 10.1061/(asce)0733-950x(2000)126:1(1).

Brivio, P. A., Giardino, C. & Zilioli, E. (2001), 'Determination of chlorophyll concentration changes in Lake Garda using an image-based radiative transfer code for Landsat TM images', *International Journal of Remote Sensing* **22**(2–3), 487–502.

Brocchini, M. (2013), 'A reasoned overview on Boussinesq-type models: the interplay between physics, mathematics and numerics', *Proceedings of the Royal Society A* **4**, 469–496.

Brommer, M. & Bochev-Van der Burgh, L. (2009), 'Sustainable coastal zone management: A concept for forecasting long-term and large-scale coastal evolution', *Journal of Coastal Research* **25**(1), 181–188.

Brown, S. A. (2016a), Numerical modelling of turbulence and sediment concentrations under breaking waves using OpenFOAM®, PhD thesis, School of Engineering, Plymouth, UK.

Brown, S. A., Greaves, D., Magar, V. & Conley, D. (2016), 'Evaluation of turbulence closure models under spilling and plunging breakers in the surf zone', *Coastal Engineering* **114**, 177–193. doi: 10.1016/j.coastaleng.2016.04.002.

Brown, S. A., Greaves, D. M., Magar, V. & Conley, D. C. (2020), 'Numerical evaluation of an implicit methodology for estimations of sediment diffusivity in wave-driven environments', Under review.

Brown, S. A., Magar, V., Greaves, D. M. & Conley, D. C. (2014), 'An evaluation of rans turbulence closure models for spilling breakers', *Coastal Engineering Proceedings* **1**(34), 5. doi: 10.9753/icce.v34.waves.5.

Bruun, P. (1954), Coastal Erosion and the Development of Beach Profiles (Beach Erosion Board Technical Memo), Technical Report 44, US Army Corps of Engineers, US Army Engineer Waterways Experiment Station, Vicksburg.

Buffault, P. (1942), *Histoire des dunes maritimes de la Gascogne*, Éditions Delmas, 1942. pp. 446. Available at: https://gallica.bnf.fr/ark:/12148/bpt6k9619289s/f8.image.texteImage. [last accessed: 01-01-2020].

Buffington, J. M. & Montgomery, D. R. (1997), 'A systematic analysis of eight decades of incipient motion studies, with special reference to gravel-bedded rivers', *Water Resources Research* **33**(8), 1993–2029. https://agupubs.onlinelibrary.wiley.com/doi/abs/10.1029/96WR03190.

Buitenkamp, M., vn den Brink, C. & van Mastrigt, A. (2016), The sand motor belongs to Everyone: The sand engine evaluation report, Anantis and Royal Haskoning. www.dezandmotor.nl/uploads/2016/09/sandmotor-policy-evaluation.pdf. [last accessed: 13-03-2019].

Cailleau, S., Fedorenko, V., Barnier, B., Blayo, E. & Debreu, L. (2008), 'Comparison of different numerical methods used to handle the open boundary of a regional ocean circulation model of the Bay of Biscay', *Ocean Modelling* **25**(1), 1–16. www.sciencedirect.com/science/article/pii/S1463500308000711.

Carling, P., Williams, J., Croudace, I. & Amos, C. (2009), 'Formation of mud ridge and runnels in the intertidal zone of the Severn Estuary, UK', *Continental Shelf Research* **29**(16), 1913–1926. doi: 10.1016/j.csr.2008.12.009.

Carriquiry, J. D., Sánchez, A. & Camacho-Ibar, V. F. (2001), 'Sedimentation in the northern Gulf of California after cessation of the Colorado River discharge', *Sedimentary Geology* **144**(1), 37–62. www.sciencedirect.com/science/article/pii/S0037073801001348.

Castagno, K. A., Jiménez-Robles, A. M., Donnelly, J. P., Wiberg, P. L., Fenster, M. S. & Fagherazzi, S. (2018), 'Intense storms increase the stability of tidal bays', *Geophysical Research Letters* **45**(11), 5491–5500. doi: 10.1029/2018gl078208.

Castedo, R., Paredes, C., de la Vega-Panizo, R. & Santos, A. P. (2017), The modelling of coastal cliffs and future trends. In: *'Hydro-Geomorphology - Models and Trends'*, InTech. doi: 10.5772/intechopen.68445.

Cerdeira-Estrada, S., Heege, T., Kolb, M., Ohlendorf, S., Uribe, A., Müller, A., Garza, R., Ressl, R., Aguirre, R., Mariño, I., Silva, R. & Martell, R. (2012), Benthic habitat and bathymetry mapping of shallow waters in Puerto morelos reefs using remote sensing with a physics based data processing, *In '2012 IEEE International Geoscience and Remote Sensing Symposium'*, pp. 4383–4386, doi:10.1109/IGARSS.2012.6350402.

Chant, R. J. (2010), *Estuarine Secondary Circulation*, Cambridge University Press, Cambridge, UK, pp. 100–124.

Chardón-Maldonado, P., Pintado-Patiño, J. C. & Puleo, J. A. (2015), 'Advances in swash-zone research: Small-scale hydrodynamic and sediment transport processes', *Coastal Engineering* **115**, 8–25.

Charru, F., Andreotti, B. & Claudin, P. (2013), 'Sand ripples and dunes', *Annual Review of Fluid Mechanics* **45**(1), 469–493. doi: 10.1146/annurev-fluid-011212-140806.

Chatzirodou, A., Karunarathna, H. & Reeve, D. E. (2016), 'Investigation of deep sea shelf sandbank dynamics driven by highly energetic tidal flows', *Marine Geology* **380**, 245–263. doi: 10.1016/j.margeo.2016.04.011.

Chatzirodou, A., Karunarathna, H. & Reeve, D. E. (2019), '3d modelling of the impacts of in-stream horizontal-axis tidal energy converters (TECs) on offshore sandbank dynamics', *Applied Ocean Research* **91**, 101882. doi: 10.1016/j.apor.2019.101882.

Chawla, A., Jay, D. A., Baptista, A. M., Wilkin, M. & Seaton, C. (2008), 'Seasonal variability and estuary–shelf interactions in circulation dynamics of a river-dominated estuary', *Estuaries and Coasts* **31**(2), 269–288. doi: 10.1007/s12237-007-9022-7.

Chen, S.-N., Geyer, W. R., Sherwood, C. R. & Ralston, D. K. (2010), 'Sediment transport and deposition on a river-dominated tidal flat: An idealized model study', *Journal of Geophysical Research: Oceans* **115**(C10), C10040. doi: 10.1029/2010JC006248.

Chernetsky, A. S., Schuttelaars, H. M. & Talke, S. A. (2010), 'The effect of tidal asymmetry and temporal settling lag on sediment trapping in tidal estuaries', *Ocean Dynamics* **60**(5), 1219–1241.

Christie, M. C. & Dyer, K. R. (1998), 'Measurements of the turbid tidal edge over the skeffling mudflats', *Geological Society, London, Special Publications* **139**(1), 45–55. doi: 10.1144/gsl.sp.1998.139.01.04.

Christie, M., Dyer, K., Blanchard, G., Cramp, A., Mitchener, H. & Paterson, D. (2000), 'Temporal and spatial distributions of moisture and organic contents across a macro-tidal mudflat', *Continental Shelf Research* **20**(10–11), 1219–1241. doi: 10.1016/s0278-4343(00)00020-0.

Christie, M., Dyer, K. & Turner, P. (1999), 'Sediment flux and bed level measurements from a macro tidal mudflat', *Estuarine, Coastal and Shelf Science* **49**(5), 667–688. www.sciencedirect.com/science/article/pii/S0272771499905255.

Cicin-Sain, B. & Belfiore, S. (2005), 'Linking marine protected areas to integrated coastal and ocean management: A review of theory and practice', *Ocean and Coastal Management* **48**(11–12), 847–868. doi: 10.1016/j.ocecoaman.2006.01.001.

Clapham, M. E. (2015), 'Unidirectional bedforms'. www.youtube.com/watch?v=mEqdT_Hmb2w.

Codiga, D. L. (2011), Analysis and prediction using the utide Matlab functions, GSO Technical Report 2011-01, Graduate School of Oceanography, University of Rhode Island, Narragansett, RI. 59 pp. www.po.gso.uri.edu/pub/downloads/codiga/pubs/2011Codiga-UTide-Report.pdf.

Collignon, A. G. & Stacey, M. T. (2013), 'Turbulence dynamics at the shoal–channel interface in a partially stratified estuary', *Journal of Physical Oceanography* **43**(5), 970–989. doi: 10.1175/jpo-d-12-0115.1.

Council, N. R. (2012), Sea-Level Rise for the Coasts of California, Oregon, and Washington: Past, Present, and Future, Technical Report, The National Academies Press, Washington, DC.

Council of Europe (1979), 'Convention on the Conservation of European Wildlife and Natural Habitats'. http://conventions.coe.int/Treaty/EN/Treaties/Html/104.htm.

Cowell, P., Roy, P. & Jones, R. (1992), 'Shoreface translation model: Computer simulation of coastal-sand-body response to sea level rise', *Mathematics and Computers in Simulation* **33**(5–6), 603–608. doi: 10.1016/0378-4754(92)90158-d.

Cowell, P., Roy, P. & Jones, R. (1995), 'Simulation of large-scale coastal change using a morphological behaviour model', *Marine Geology* **126**(1–4), 45–61. doi: 10.1016/0025-3227(95)00065-7.

Cowell, P. J., Stive, M. J. F., Niedoroda, A. H., de Vriend, H. J., Swift, D. J. P., Kaminsky, G. M. & Capobianco, M. (2003), 'The coastal-tract (part 1): A conceptual approach to aggregated modeling of low-order coastal change', *Jouranl of Coastal Reserch* **19**(4), 812–827.

Cuadra, L., Salcedo-Sanz, S., Nieto-Borge, J., Alexandre, E. & Rodríguez, G. (2016), 'Computational intelligence in wave energy: Comprehensive review and case study', *Renewable and Sustainable Energy Reviews* **58**, 1223–1246. doi: 10.1016/j.rser.2015.12.253.

D'Alpaos, A., Lanzoni, S., Marani, M., Fagherazzi, S. & Rinaldo, A. (2005), 'Tidal network ontogeny: Channel initiation and early development', *Journal of Geophysical Research: Earth Surface* **110**(F2). https://agupubs.onlinelibrary.wiley.com/doi/abs/10.1029/2004JF000182.

D'Alpaos, A., Lanzoni, S., Marani, M. & Rinaldo, A. (2007), 'Landscape evolution in tidal embayments: Modeling the interplay of erosion, sedimentation, and vegetation dynamics', *Journal of Geophysical Research: Earth Surface* **112**(F1). https://agupubs.onlinelibrary.wiley.com/doi/abs/10.1029/2006JF000537.

Darrigol, O. (2005), *Worlds of Flow: A History of Hydrodynamics from the Bernoullis to Prandtl*, Oxford University Press, Oxford.

Davies, A. G. & Thorne, P. D. (2008), 'Advances in the study of moving sediments and evolving seabeds', *Surveys in Geophysics* **29**(1), 1–36. doi: 10.1007/s10712-008-9039-x.

Davies, A. G. & Robins, P. (2017), 'Residual flow, bedforms and sediment transport in a tidal channel modelled with variable bed roughness', *Geomorphology* **295**, 855–872. doi: 10.1016/j.geomorph.2017.08.029.

de Goede, E. D., Groeneweg, J., Tan, K. H., Borsboom, M. J. A. & Stelling, G. S. (1995), 'A domain decomposition method for the three-dimensional shallow water equations', *Simulation Practice and Theory* **3**, 307–325.

de Schipper, M. A., de Vries, S., Ruessink, G., de Zeeuw, R. C., Rutten, J., van Gelder-Maas, C. & Stive, M. J. (2016), 'Initial spreading of a mega feeder nourishment: Observations of the sand engine pilot project', *Coastal Engineering* **111**, 23–38. doi: 10.1016/j.coastaleng.2015.10.011.

de Vriend, H., Capobianco, M., Chesher, T., de Swart, H., Latteux, B. & Stive, M. (1993), 'Approaches to long-term modelling of coastal morphology:

A review', *Coastal Engineering* **21**(1), 225–269. Special Issue Coastal Morphodynamics: Processes and Modelling. www.sciencedirect.com/science/article/pii/0378383993900519.

de Vriend, H. J. (1991), 'Mathematical modelling and large-scale coastal behaviour', *Journal of Hydraulic Research* **29**(6), 727–740.

Dean, R. (1973), Heuristic models of sand transport in the surf zone, *In Proceeding of the Conference on Engineering Dynamics in the Surf Zone*, Sydney, Australia, May, 1973, First Australian Coastal Engineering Conference. pp. 208–214.

Dean, R. G. (1991), 'Equilibrium beach profiles: Characteristics and applications', *Journal of Coastal Research* **7**(1), 53–84.

Dean, R. G. & Dalrymple, R. A. (1984), *Water Wave Mechanics for Engineers and Scientists*, Prentice-Hall, New York.

Defant, A. (1958), *Ebb and Flow: The Tides of Earth, Air and Water*, The University of Michigan Press, Ann Arbor, MI.

Defne, Z., Haas, K. A. & Fritz, H. M. (2011), 'Numerical modeling of tidal currents and the effects of power extraction on estuarine hydrodynamics along the Georgia coast, USA', *Renewable Energy* **36**(12), 3461–3471. doi: 10.1016/j.renene.2011.05.027.

DEFRA (2006), Shoreline management plan guidance volume 1: Aims and requirements, Technical Report, Department for Environment, Food and Rural Affairs (DEFRA), London, UK.

Deltares (2018), Delft3D-FLOW User Manual Version 3.15: Simulation of multi-dimensional hydrodynamic flows and transport phenomena, including sediments, Technical Report, Deltares.

Depietri, Y., Dahal, K. & McPhearson, T. (2018), 'Multi-hazard risks in New York City', *Natural Hazards and Earth System Sciences* **18**(12), 3363–3381. doi: 10.5194/nhess-18-3363-2018.

Dijkstra, Y. M., Schuttelaars, H. M. & Burchard, H. (2017), 'Generation of exchange flows in estuaries by tidal and gravitational eddy viscosity-shear covariance (ESCO)', *Journal of Geophysical Research: Oceans* **122**(5), 4217–4237.

Dohmen-Janssen, C. & Hanes, D. (2002), 'Sheet flow dynamics under monochromatic nonbreaking waves', *Journal of Geophysical Research: Oceans* **107**(C10, 3149).

Dohmen-Janssen, C. & Hanes, D. (2005), 'Sheet flow and suspended sediment due to wave groups in a large wave flume', *Continental Shelf Research* **25**(3), 333–347.

Dohmen-Janssen, C., Hassan, W. & Ribberink, J. (2001), 'Mobile-bed effects in oscillatory sheet flow', *Journal of Geophysical Research: Oceans* **106**(C11), 27103–27115.

Doody, J. P. (2013), 'Coastal squeeze and managed realignment in southeast England, does it tell us anything about the future?' *Ocean and Coastal Management* **79**, 34–41. Managing Estuarine Sediments. www.sciencedirect.com/science/article/pii/S0964569112001172.

Doxaran, D., Froidefond, J. M., Castaing, P. & Barin, M. (2009), 'Dynamics of the turbidity maximum zone in a macrotidal estuary (the Gironde, France): Observations from field and MODIS satellite data', *Estuarine, Coastal and Shelf Science* **81**, 321–332.

Dyer, K. R. (1982), 'The initiation of sedimentary furrows by standing internal waves', *Sedimentology* **29**(6), 885–889. doi: 10.1111/j.1365-3091.1982.tb00091.x.

Dyer, K. R. (1997), *Estuaries: A Physical Introduction*, 2nd edn, John Wiley and Son, Chichester.

Egbert, G. D., Bennett, A. F. & Foreman, M. G. G. (1994), 'TOPEX/ POSEIDON tides estimated using a global inverse model', *Journal of Geophysical Research* **99**(C12), 24821. doi: 10.1029/94jc01894.

Egbert, G. D. & Erofeeva, S. Y. (2002), 'Efficient inverse modeling of barotropic ocean tides', *Journal of Atmospheric and Oceanic Technology* **19**(2), 183–204. doi: 10.1175/1520-0426(2002)019<0183:eimobo>2.0.co;2.

Einstein, H. A. (1950), The bed-load function for sediment transportation in open-channel flows, Technical Report 1026, U.S. Department of Agriculture, Soil Conservation Service, Technical Bulletin.

Ekman, M. (1993), 'A concise history of the theories of tides, precession-nutation and polar motion (from antiquity to 1950)', *Surveys in Geophysics* **14**(6), 585–617. doi: 10.1007/bf00666031.

Elfrink, B., Hanes, D. M. & Ruessink, B. G. (2006), 'Parametrization and simulation of near bed orbital velocities under irregular waves in shallow water', *Coastal Engineeering* **53**, 915–927.

Elgar, S. & Guza, R. T. (1985), 'Shoaling gravity waves: Comparisons between field observations, linear theory, and a nonlinear model', *Journal of Fluid Mechanics* **158**(1), 47. doi: 10.1017/s0022112085002543.

Elias, E. P. L., Gelfenbaum, G. & Van der Westhuysen, A. J. (2012), 'Validation of a coupled wave-flow model in a high-energy setting: The mouth of the Columbia River', *Journal of Geophysical Research: Oceans* **117**(C9). doi: 10.1029/2012jc008105.

Elliott, M., Nedwell, S., Jones, N. V., Read, S. J., Cutts, N. D. & Hemingway, K. L. (1998), Intertidal sand and mudflats and subtidal mobile sandbanks (volume II). An overview of dynamic and sensitivity characteristics for conservation management of marine SACs, Technical Report II, Scottish Association for Marine Science (UK Marine SACs Project).

Elmilady, H., Van der Wegen, M., Roelvink, D. & Jaffe, B. E. (2018), 'Intertidal area disappears under sea level rise: 250 years of morphodynamic modelling in San Pablo Bay, California', *Journal of Geophysical Research: Earth Surface*. doi: 10.1029/2018jf004857.

Elsner, J. B. & Tsonis, A. A. (1996), *Singular Spectrum Analysis*, Springer, New York. doi: 10.1007/978-1-4757-2514-8.

Engelund, F. & Fredsøe, J. (1976), 'A sediment transport model for straight alluvial channels', *Hydrology Research* **7**(5), 293–306. doi: 10.2166/nh.1976.0019.

Esteves, L. S., ed. (2014), *Managed Realignment: A Viable Long-Term Coastal Management Strategy?* Springer Briefs in Environmental Science, 1st edn, Springer, Dordrecht, the Netherlands.

Exner, F. M. (1920), 'Zur physik der dünen', *Akademie der Wissenschaften in Wien, Sitzungsberichte, Mathematisch-naturwissenschaftliche Klasse, Abteilung* **129**(2a), 929–952.

Exner, F. M. (1925), 'Über die wechselwirkung zwischen wasser und geschiebe in flüssen', *Akad. Wiss. Wien Math. Naturwiss. Klasse* **134**(2a), 165–204.

Fagherazzi, S. & Furbish, D. J. (2001), 'On the shape and widening of salt marsh creeks', *Journal of Geophysical Research: Oceans* **106**(C1), 991–1003. https://agupubs.onlinelibrary.wiley.com/doi/abs/10.1029/1999JC000115.

Fagherazzi, S. & Mariotti, G. (2012), 'Mudflat runnels: Evidence and importance of very shallow flows in intertidal morphodynamics', *Geophysical Research Letters* **39**(14). doi: 10.1029/2012gl052542.

Fairley, I., Davidson, M., Kingston, K., Dolphin, T. & Phillips, R. (2009), 'Empirical orthogonal function analysis of shoreline changes behind two different designs of detached breakwaters', *Coastal Engineering* **56**(11–12), 1097–1108. doi: 10.1016/j.coastaleng.2009.08.001.

Falqués, A. (2003), 'On the diffusivity in coastline dynamics', *Geophysical Research Letters* **30**(21). https://agupubs.onlinelibrary.wiley.com/doi/abs/10.1029/2003GL017760.

Falqués, A. & Calvete, D. (2004), 'Large-scale dynamics of sandy coastlines: Diffusivity and instability', *Journal of Geophysical Research: Oceans* **110**(C3). https://agupubs.onlinelibrary.wiley.com/doi/abs/10.1029/2004JC002587.

Feddersen, F., Gallagher, E., Guza, R. & Elgar, S. (2003), 'The drag coefficient, bottom roughness, and wave-breaking in the nearshore', *Coastal Engineering* **48**(3), 189–195. doi: 10.1016/s0378-3839(03)00026-7.

FEMA (2015), Guidance for flood risk analysis and mapping: coastal erosion, Technical Report, Federal Emergency Management Agency (FEMA), Washington, DC.

Feng, S., Cheng, R. T. & Pangen, X. (1986), 'On tide-induced Lagrangian residual current and residual transport: 1. Lagrangian residual current', *Water Resources Research* **22**(12), 1623–1634. doi: 10.1029/WR022i012p01623.

Fields, P. (2015), 'Chapter 9 tides'. https://slideplayer.com/user/5532010/.

Fisher, C. M., Young, G. S., Winstead, N. S. & Haqq-Misra, J. D. (2008), 'Comparison of synthetic aperture radar–derived wind speeds with buoy wind speeds along the mountainous Alaskan coast', *Journal of Applied Meteorology and Climatology* **47**(5), 1365–1376. doi: 10.1175/2007jamc1716.1.

Flather, R. A. (1984), 'A numerical model investigation of the storm surge of 31 January and 1 February 1953 in the North Sea', *Quarterly Journal of the Royal Meteorological Society* **110**(465), 591–612. https://rmets.onlinelibrary.wiley.com/doi/abs/10.1002/qj.49711046503.

Folk, R. L. (1954), 'The distinction between grain size and mineral composition in sedimentary-rock nomenclature', *The Journal of Geology* **62**(4), 344–359.

Foster, D. L., Bowen, A. J., Holman, R. A. & Natoo, P. (2006), 'Field evidence of pressure gradient induced incipient motion', *Journal of Geophysical Research: Oceans* **111**(C5). https://agupubs.onlinelibrary.wiley.com/doi/abs/10.1029/2004JC002863.

Fountain, L., Sexton, J., Habili, N., Hazelwood, M. & Anderson, H. (2010), Storm surge modelling for Bunbury, Western Australia, Technical Report 2010/04, Autralian Government Geoscience Australia. Professional Opinion.

Frandsen, J. B. (2004), 'Sloshing motions in excited tanks', *Journal of Computational Physics* **196**(1), 53–87. doi: 10.1016/j.jcp.2003.10.031.

Fredsøe, J. & Deigaard, R. (1992), *Mechanics of Coastal Sediment Transport*, World Scientific. doi: 10.1142/1546.

French, P. (1997), *Coastal and Estuarine Management*, Routledge Environmental Management Series, Routledge. https://books.google.com.mx/books?id=q4BllAb_PeoC.

Frey, P. J. & George, P.-L. (2008), *Mesh Generation*, ISTE. doi: 10.1002/9780470611166.

Friedrichs, C. T. & Aubrey, D. G. (2013), *Uniform Bottom Shear Stress and Equilibrium Hyposometry of Intertidal Flats*, American Geophysical Union, pp. 405–429. doi: 10.1029/CE050p0405.

Fugate, D. C., Friedrichs, C. T. & Sanford, L. P. (2007), 'Lateral dynamics and associated transport of sediment in the upper reaches of a partially mixed estuary, Chesapeake Bay, USA', *Continental Shelf Research* **27**(5), 679–698. www.sciencedirect.com/science/article/pii/S0278434306003682.

Funaro, D., Quarteroni, A. & Zanolli, P. (1988), 'An iterative procedure with interface relaxation for domain decomposition methods', *SIAM Journal on Numerical Analysis* **25**(6), 1213–1236.

Ganju, N. K., Knowles, N. & Schoellhamer, D. H. (2008), 'Temporal downscaling of decadal sediment load estimates to a daily interval for use in hindcast simulations', *Journal of Hydrology* **349**(3–4), 512–523. doi: 10.1016/j.jhydrol.2007.11.026.

Ganju, N. K., Schoellhamer, D. H. & Jaffe, B. E. (2009), 'Hindcasting of decadal-timescale estuarine bathymetric change with a tidal-timescale model', *Journal of Geophysical Research* **114**(F4). doi: 10.1029/2008jf 001191.

García-Hermosa, M. I., Borthwick, A. G. L., Soulsby, R. L., Stansby, P. K., Taylor, P. H., Huang, J. & Zhou, J. G. (2009), Interpretation of large-scale morphodynamic laboratory experiments: Spoil heaps and sandbanks. In: J. M. Smith (ed.) *'Coastal Engineering 2008'*, World Scientific Publishing Company, Singapore, pp. 2609-2621.

Gelfenbaum, G. & Kaminsky, G. M. (2010), 'Large-scale coastal change in the Columbia River littoral cell: An overview', *Marine Geology* **273**(1–4), 1–10. doi: 10.1016/j.margeo.2010.02.007.

Geo (2019), ANUGA User Manual Release 2.0.3.

Geyer, W. R. (2010), *Estuarine Salinity Structure and Circulation*, Cambridge University Press, Cambridge, UK, pp. 12–26.

Ghil, M. (2002), 'Advanced spectral methods for climatic time series', *Reviews of Geophysics* **40**(1). doi: 10.1029/2000rg000092.

Giardino, C., Candiani, G. & Zilioli, E. (2005), 'Detecting chlorophyll-a in Lake Garda using TOA MERIS radiances', *Photogrammetric Engineering and Remote Sensing* **71**, 1045–1051.

Giese, B. S. & Jay, D. A. (1989), 'Modelling tidal energetics of the Columbia River Estuary', *Estuarine, Coastal and Shelf Science* **29**(6), 549–571. doi: 10.1016/0272-7714(89)90010-3.

Gitelson, A. A., Gurlin, D., Moses, W. J. & Barrow, T. (2009), 'A bio-optical algorithm for the remote estimation of the chlorophyll- a concentration in case 2 waters', *Environmental Research Letters* **4**(4), 045003. http://stacks.iop.org/1748-9326/4/i=4/a=045003.

Glavovic, B. C., Kelly, M., Kay, R. & Travers, A., (eds) (2015), *Climate Change and the Coast: Building Resilient Communities*, CRC Press, Boca Raton, FL.

Goda, Y. (1970), 'Numerical experiments on wave statistics with spectral simulation', *Report of Port and Harbour Research Institute* **9**(3), 3–57.

Goda, Y. (2010), *Random Seas and Design of Maritime Structures*, World Scientific. doi: 10.1142/7425.

Godin, G. (1972), *The Analysis of Tides*, Liverpool University Press. https://books.google.com.mx/books?id=rxUIAQAAIAAJ.

Goff, J. R., Lane, E. & Arnold, J. (2009), 'The tsunami geomorphology of coastal dunes', *Natural Hazards and Earth System Science* **9**(3), 847–854. doi: 10.5194/nhess-9-847-2009.

Gross, M. S. & Magar, V. (2015), 'Offshore wind energy potential estimation using UPSCALE climate data', *Energy Science and Engineering* **3**(4), 342–359. doi: 10.1002/ese3.76.

Gruwez, V., Verheyen, B., Wauters, P. & Bolle, A. (2016), 'Hindcasting sand spit morphodynamics after groyne construction in Ghana', *Journal of Applied Water Engineering and Research* **5**(2), 167–176. doi: 10.1080/23249676.2016.1184597.

Hahmann, A. N., Vincent, C. L., Peña, A., Lange, J. & Hasager, C. B. (2014), 'Wind climate estimation using WRF model output: method and model sensitivities over the sea', *International Journal of Climatology* **35**(12), 3422–3439. doi: 10.1002/joc.4217.

Hamm, L., Madsen, P. A. & Peregrine, D. (1993), 'Wave transformation in the nearshore zone: A review', *Coastal Engineering* **21**(1–3), 5–39. doi: 10.1016/0378-3839(93)90044-9.

Hansen, W. (1956), 'Theorie zur errechnung des wasserstandes und der strömungen in randmeeren nebst anwendungen', *Tellus* **8**(3), 287–300.

Hanson, H. & Larson, M. (2000), *Simulating Coastal Evolution Using a New Type of N-Line Model*, American Society of Civil Engineers (ASCE), Reston, VA, pp. 2808–2821.

Hasselmann, K. (1966), 'Feynman diagrams and interaction rules of wave-wave scattering processes', *Reviews of Geophysics* **4**(1), 1. doi: 10.1029/rg004i001p00001.

Hasselmann, K., Barnett, T., Bouws, E., Carlson, H., Cartwright, D., Enke, K., Ewing, J., Gienapp, H., Hasselmann, D., Kruseman, P., Meerburg, A., Müller, P., Olbers, D. J., Richter, K., Sell, W. & Walden, H. (1973), Measurements of wind-wave growth and swell decay during the joint north sea wave project (JONSWAP), Technical Report 8 (12), Deutsches Hydrographisches Institut, Hamburg.

Hattori, M. & Kawamata, R. (1980), 'Onshore-offshore transport and beach profile change', *Coastal Engineering Proceedings* **17**(25). https://icce-ojs-tamu.tdl.org/icce/index.php/icce/article/view/3495.

Hequette, A. & Aernouts, D. (2010), 'The influence of nearshore sand bank dynamics on shoreline evolution in a macrotidal coastal environment, Calais, northern France', *Continental Shelf Research* **30**(12), 1349–1361. doi: 10.1016/j.csr.2010.04.017.

Herbers, T. H. C. (2003), 'Shoaling transformation of wave frequency-directional spectra', *Journal of Geophysical Research* **108**(C1). doi: 10.1029/2001jc001304.

Herdman, L., Erikson, L. & Barnard, P. (2018), 'Storm surge propagation and flooding in small tidal rivers during events of mixed coastal and fluvial influence', *Journal of Marine Science and Engineering* **6**(4), 158. doi: 10.3390/jmse6040158.

Hibma, A., Stive, M. & Wang, Z. (2004), 'Estuarine morphodynamics', *Coastal Engineering* **51**(8), 765–778. Coastal Morphodynamic Modeling. www.sciencedirect.com/science/article/pii/S037838390400081X.

Hir, P. L., Cayocca, F. & Waeles, B. (2011), 'Dynamics of sand and mud mixtures: A multiprocess-based modelling strategy', *Continental Shelf Research* **31**(10), S135–S149. doi: 10.1016/j.csr.2010.12.009.

Hir, P. L., Monbet, Y. & Orvain, F. (2007), 'Sediment erodability in sediment transport modelling: Can we account for biota effects?' *Continental Shelf Research* **27**(8), 1116–1142. doi: 10.1016/j.csr.2005.11.016.

Hjulstroṁ, F. (1935), Studies of the morphological activity of rivers as illustrated by the River Fyris, PhD thesis, Uppsala Universitet, Uppsala, Sweden. http://catalog.hathitrust.org/api/volumes/oclc/8108973.html.

Hoefel, F. (2003), 'Wave-induced sediment transport and sandbar migration', *Science* **299**(5614), 1885–1887. doi: 10.1126/science.1081448.

Holland, G. J. (1980), 'An analytic model of the wind and pressure profiles in hurricanes', *Monthly Weather Review* **108**(8), 1212–1218. doi: 10.1175/1520-0493(1980)108<1212:aamotw>2.0.co;2.

Holthuijsen, L. H. (2007), *Waves in Oceanic and Coastal Waters*, Cambridge University Press, Cambridge.

Hoover, J. D., Stauffer, D. R., Richardson, S. J., Mahrt, L., Gaudet, B. J. & Suarez, A. (2015), 'Submeso motions within the stable boundary layer and their relationships to local indicators and synoptic regime in moderately complex terrain', *Journal of Applied Meteorology and Climatology* **54**, 352–369.

Hosegood, P. J., Gregg, M. C. & Alford, M. H. (2008), 'Restratification of the surface mixed layer with submesoscale lateral density gradients: Diagnosing the importance of the horizontal dimension', *Journal of Physical Oceanography* **38**(11), 2438–2460. doi: 10.1175/2008jpo3843.1.

Hsu, T.-J., & Hanes, D.M. (2004), 'Effects of wave shape on sheet flow sediment transport', *Journal of Geophysical Research* **109**, (C05025).

Hsu, T.-J., Chen, S.-N. & Ogston, A. S. (2013), 'The landward and seaward mechanisms of fine-sediment transport across intertidal flats in the shallow-water region—a numerical investigation', *Continental Shelf Research* **60**, S85–S98. Hydrodynamics and sedimentation on mesotidal sand- and mudflats. www.sciencedirect.com/science/article/pii/S0278434312000313.

Hubbert, G. D., Leslie, L. M. & Manton, M. J. (1990), 'A storm surge model for the Australian region', *Quarterly Journal of the Royal Meteorological Society* **116**(494), 1005–1020. doi: 10.1002/qj.49711649411.

Hubbert, G. & McInnes, K. (1999), 'A storm surge inundation model for coastal planning and impact studies', *Journal of Coastal Research*, **15** 168–85.

Huijts, K. M. H., Schuttelaars, H. M., de Swart, H. E. & Valle-Levinson, A. (2006), 'Lateral entrapment of sediment in tidal estuaries: An idealized model study', *Journal of Geophysical Research: Oceans* **111**(C12), C12016. doi: 10.1029/2006JC003615.

Hunt-Raby, A., Othman, I. K., Jayaratne, R., Bullock, G. & Bredmose, H. (2010), Effect of protruding roughness elements on wave overtopping. In *'Coasts, Marine Structures and Breakwaters: Adapting to Change'*, Thomas Telford Ltd, pp. 574–586. doi: 10.1680/cmsb.41318.0054.

Huthnance, J. (1973), 'Tidal current asymmetries over the Norfolk Sandbanks', *Estuarine and Coastal Marine Science* **1**(1), 89–99. www.sciencedirect.com/science/article/pii/0302352473900613.

Iglesias, G., Tercero, J. A., Simas, T., Machado, I. & Cruz, E. (2018), Environmental effects. In: *'Wave and Tidal Energy'*, John Wiley & Sons, Ltd, pp. 364–454. doi: 10.1002/9781119014492.ch9.

Instream Energy Systems (2018). www.instreamenergy.com/. [accessed: 25-12-2018]

Irish, J. L. & Cañizares, R. (2009), 'Storm-wave flow through tidal inlets and its influence on bay flooding', *Journal of Waterway, Port, Coastal, and Ocean Engineering* **135**(2), 52–60.

Jacobs, C. T. & Piggott, M. D. (2015), 'Firedrake-Fluids v0.1: numerical modelling of shallow water flows using an automated solution framework', *Geoscientific Model Development* **8**(3), 533–547. www.geosci-model-dev.net/8/533/2015/.

Jacobsen, N. G. (2011), A Full Hydro- and Morphodynamic Description of Breaker Bar Development, PhD thesis, Technical University of Denmark, Department of Mechanical Engineering. DCAMM Special Report no. S136.

Jacobsen, N. G. & Fredsøe, J. (2014), 'Formation and development of a breaker bar under regular waves. Part 2: Sediment transport and morphology', *Coastal Engineering* **88**, 55–68. doi: 10.1016/j.coastaleng.2014.01.015.

Jacobsen, N. G., Fredsøe, J. & Jensen, J. H. (2014), 'Formation and development of a breaker bar under regular waves. Part 1: Model description and hydrodynamics', *Coastal Engineering* **88**, 182–193. doi: 10.1016/j.coastaleng.2013.12.008.

Jamal, M., Simmonds, D. & Magar, V. (2014), 'Modelling gravel beach dynamics with XBeach', *Coastal Engineering* **89**, 20–29. doi: 10.1016/j.coastaleng.2014.03.006.

Jamal, M. H., Simmonds, D. J. & Magar, V. (2012), 'Gravel beach profile evoloution in wave and tidal environments', *Coastal Engineering Proceedings* **1**(33), 15. doi: 10.9753/icce.v33.sediment.15.

Janssen, T. T., Herbers, T. H. C. & Battjes, J. A. (2008), 'Evolution of ocean wave statistics in shallow water: Refraction and diffraction over seafloor topography', *Journal of Geophysical Research* **113**(C3). doi: 10.1029/2007jc004410.

Jay, D. A. (2010), *Estuarine Variability*, Cambridge University Press, Cambridge, UK, pp. 62–99.

Jayaratne, R., Abimola, A., Mikami, T., Matsuba, S., Esteban, M. & Shibayama, T. (2014), 'Predictive model for scour depth of coastal structure failures due to tsunamis', *Coastal Engineering Proceedings* **1**(34), 56. doi: 10.9753/icce.v34.structures.56.

Kamphuis, J. (2000), *Introduction to Coastal Engineering and Management*, Advanced series on ocean engineering, World Scientific. https://books.google.com.mx/books?id=XZhwW8md8JYC.

Kamphuis, J. W. (2010), *Introduction to Coastal Engineering and Management*, World Scientific. doi: 10.1142/7021.

Kelly, P. M. (2015), *Climate Drivers in the Coastal Zone*, CRC Press, Boca Raton, FL, Chapter 2.

Kerin, I., Giri, S. & Bekić, D. (2018), Simulation of levee breach using delft models: A case study of the Drava River flood event. In:*'Advances in Hydroinformatics'*, Springer, Singapore, pp. 1117–1131. doi: 10.1007/978-981-10-7218-5_77.

Kimmerer, W. J. (2004), 'Open water processes of the San Francisco estuary: From physical forcing to biological responses', *San Francisco Estuary and Watershed Science* **2**(1). doi: 10.15447/sfews.2004v2iss1art1.

Kirwan, M. L., Guntenspergen, G. R., D'Alpaos, A., Morris, J. T., Mudd, S. M. & Temmerman, S. (2010), 'Limits on the adaptability of coastal marshes to rising sea level', *Geophysical Research Letters* **37**(23). doi: 10.1029/2010gl045489.

Kirwan, M. L. & Megonigal, J. P. (2013), 'Tidal wetland stability in the face of human impacts and sea-level rise', *Nature* **504**(7478), 53–60. doi: 10.1038/nature12856.

Kirwan, M. L. & Murray, A. B. (2007), 'A coupled geomorphic and ecological model of tidal marsh evolution', *Proceedings of the National Academy of Sciences* **104**(15), 6118–6122. doi: 10.1073/pnas.0700958104.

Kono, T., McCall, R. & Magar, V. (2018), 'Analysis of coastal erosion and flooding protection provided by a Caribbean coral reef during hurricane Wilma'.

Kono, T., McCall, R. & Magar, V. (2019), 'Analysis of coastal erosion and flooding protection provided by a Caribbean coral reef during Hurricane Wilma'. Under Review.

Kullenberg, G. E. B. (1976), 'On vertical mixing and the energy transfer from the wind to the water', *Tellus* **28**(2), 159–165.

Kutser, T., Pierson, D. C., Kallio, K. Y., Reinart, A. & Sobek, S. (2005), 'Mapping lake CDOM by satellite remote sensing', *Remote Sensing of Environment* **94**(4), 535–540. www.sciencedirect.com/science/article/pii/S0034425704003670.

Lanckriet, T. & Puleo, J. A. (2015), 'A semianalytical model for sheet flow layer thickness with application to the swash zone', *Journal of Geophysical Research: Oceans* **120**(2), 1333–1352.

Larson, M., Capobianco, M., Jansen, H., Rozynski, G., Southgate, H., Stive, M., Wijnberg, K. & Hulscher, S. (2003), 'Analysis and modeling of field data of coastal morphological evolution over yearly and decadal time scales. Part 1: Background and linear techniques', *Journal of Coastal Research* **19**(4), 760–775.

Latham, J.-P., Xiang, J., Anastasaki, E., Guo, L., Karantzoulis, N., Vire, A. & Pain, C. (2014), 'Numerical modelling of forces, stresses and breakages of concrete armour units', *Coastal Engineering Proceedings* **1**(34), 78. doi: 10.9753/icce.v34.structures.78.

Launder, B. E., Reece, G. J. & Rodi, W. (1975), 'Progress in the development of a Reynolds-stress turbulence closure', *Journal of Fluid Mechanics* **68**(03), 537. doi: 10.1017/s0022112075001814.

Lazure, P. & Dumas, F. (2008), 'An external–internal mode coupling for a 3D hydrodynamical model for applications at regional scale (MARS)', *Advances in Water Resources* **31**(2), 233–250. doi: 10.1016/j.advwatres.2007.06.010.

Lefebvre, A. & Winter, C. (2016), 'Predicting bed form roughness: The influence of lee side angle', *Geo-Marine Letters* **36**(2), 121–133.

Leggett, D., Cooper, N. & Harvey, R. (2004), Coastal and estuarine managed realignment, Technical Report, CIRIA, London, UK.

Leslie, L. M., Mills, G. A., Logan, L. W., Gauntlett, D. J., Kelly, G. A., Manton, M. J., McGregor, J. L. & Sardie, J. M. (1985), 'A high resolution primitive equations NWP model for operations and research', *Australian Meteorological Magazine* **33**(1), 11–35.

Lesser, G., Roelvink, J., Van Kester, J. & Stelling, G. (2004), 'Development and validation of a three-dimensional morphological model', *Coastal Engineering* **51**(8–9), 883–915. doi: 10.1016/j.coastaleng.2004.07.014.

Li, C. & Valle-Levinson, A. (1999), 'A two-dimensional analytic tidal model for a narrow estuary of arbitrary lateral depth variation: The intratidal motion', *Journal of Geophysical Research: Oceans* **104**(C10), 23525–23543.

Li, M. Z., Hannah, C. G., Perrie, W. A., Tang, C. C., Prescott, R. H. & Greenberg, D. A. (2015), 'Modelling seabed shear stress, sediment mobility, and sediment transport in the Bay of Fundy', *Canadian Journal of Earth Sciences* **52**(9), 757–775.

Lin, J. & Kuo, A. Y. (2001), 'Secondary turbidity maximum in a partially mixed microtidal estuary', *Estuaries* **24**(5), 707–720.

Locke, J. L. (1971), Sedimentation and foraminiferal aspects of the recent sediments of San Pablo Bay, Master's thesis, Department of Geology, San Jose State College, San Jose, CA.

Lorke, A., Umlauf, L., Jonas, T. & Wüest, A. (2002), 'Dynamics of turbulence in low-speed oscillating bottom-boundary layers of stratified basins', *Environmental Fluid Mechanics* **2**, 291–313.

Lowe, J. A., Gregory, J. M. & Flather, R. A. (2001), 'Changes in the occurrence of storm surges around the united kingdom under a future climate scenario using a dynamic storm surge model driven by the hadley centre climate models', *Climate Dynamics* **18**(3), 179–188.

Lowe, J. A., Woodworth, P. L., Knutson, T., McDonald, R. E., McInnes, K. L., Woth, K., von Storch, H., Wolf, J., Swail, V., Bernier, N. B., Gulev, S., Horsburgh, K. J., Unnikrishnan, A. S., Hunter, J. R. & Weisse, R. (2010), *Past and Future Changes in Extreme Sea Levels and Waves*, Wiley-Blackwell, Hoboken, NJ, pp. 326–375.

Lowe, R. J., Falter, J. L., Koseff, J. R., Monismith, S. G. & Atkinson, M. J. (2007), 'Spectral wave flow attenuation within submerged canopies: Implications for wave energy dissipation', *Journal of Geophysical Research* **112**(C5). doi: 10.1029/2006jc003605.

Longuet-Higgins, M. S. (1953), 'Mass transport in water waves', *Philosophical Transactions of the Royal Society of London A: Mathematical, Physical and Engineering Sciences* **245**(903), 535–581. http://rsta.royalsocietypublishing.org/content/245/903/535.

Luijendijk, A. P., Ranasinghe, R., de Schipper, M. A., Huisman, B. A., Swinkels, C. M., Walstra, D. J. R. & Stive, M. J. F. (2017), 'The initial morphological response of the Sand Engine: A process-based modelling study', *Coastal Engineering* **119**, 1–14.

MacCready, P. & Geyer, W. R. (2010), 'Advances in Estuarine Physics', *Annual Review of Marine Science* **2**(1), 35–58. www.annualreviews.org/doi/10.1146/annurev-marine-120308-081015.

MacCready, P. & Geyer, W. R. (2014), 'The estuarine circulation', *Annual Review of Fluid Mechanics* **46**, 175–197.

MacVean, L. J. & Lacy, J. R. (2014), 'Interactions between waves, sediment, and turbulence on a shallow estuarine mudflat', *Journal of Geophysical Research: Oceans* **119**(3), 1534–1553. doi: 10.1002/2013jc009477.

Madsen, O. S. & Grant, W. D. (1977), Quantitative description of sediment transport by waves, In: *'Coastal Engineering 1976'*, American Society of Civil Engineers. doi: 10.1061/9780872620834.065.

Madsen, P. A. & Sørensen, O. R. (1992), 'A new form of the Boussinesq equations with improved linear dispersion characteristics. Part 2. A slowly-varying bathymetry', *Coastal Engineering* **18**(3–4), 183–204. doi: 10.1016/0378-3839(92)90019-q.

Magar, V. (2008), 'Behaviour-based models'. www.coastalwiki.org/wiki/Behaviour-based_models.

Magar, V. (2016), *Estuarine Sediment Composition*, Springer, Dordrecht, Netherlands, pp. 285–289.

Magar, V. (2018), Tidal current technologies. In: *'Sustainable Energy Technologies'*, CRC Press, pp. 293–308. doi: 10.1201/9781315269979-18.

Magar, V., Gross, M. & González-García, L. (2018), 'Offshore wind energy resource assessment under techno-economic and social-ecological constraints', *Ocean and Coastal Management* **152**, 77–87. doi: 10.1016/j.ocecoaman.2017.10.007.

Magar, V., Lefranc, M., Hoyle, R. B. & Reeve, D. E. (2012), 'Spectral quantification of nonlinear behaviour of the nearshore seabed and correlations with potential forcings at Duck, N.C., U.S.A.', *PLoS One* **7**(6), e39196. doi: 10.1371/journal.pone.0039196.

Malarkey, J. & Davies, A. (2002), 'Discrete vortex modelling of oscillatory flow over ripples', *Applied Ocean Research* **24**(3), 127–145. doi: 10.1016/s0141-1187(02)00035-4.

Malarkey, J., Magar, V. & Davies, A. (2015), 'Mixing efficiency of sediment and momentum above rippled beds under oscillatory flows', *Continental Shelf Research* **108**, 76–88. doi: 10.1016/j.csr.2015.08.004.

Mangor, K., Drønen, N. K., Kaergaard, K. H. & Kristensen, S. E. (2017), Shoreline management guidelines, Technical Report, Danish Hydraulic Institute, Hørsholm, Denmark.

Marcello, J., Eugenio, F., Estrada-Allis, S. & Sangrà, P. (2015), 'Segmentation and tracking of anticyclonic eddies during a submarine volcanic eruption using ocean colour imagery', *Sensors* **15**(4), 8732–8748. doi: 10.3390/s150408732.

Mariotti, G. & Fagherazzi, S. (2010), 'A numerical model for the coupled long-term evolution of salt marshes and tidal flats', *Journal of Geophysical Research* **115**(F1). doi: 10.1029/2009jf001326.

Mariotti, G. & Fagherazzi, S. (2012), 'Channels-tidal flat sediment exchange: The channel spillover mechanism', *Journal of Geophysical Research: Oceans* **117**(C3). doi: 10.1029/2011jc007378.

Martin, J. (2018), 'Underwater? Study says thousands of Florida homes at risk of chronic inundation by 2045', *The Saint Augustine Record*. www.staugustine.com/news/20180618/underwater-study-says-thousands-of-florida-homes-at-risk-of-chronic-inundation-by-2045.

Martínez, M., Intralawan, A., Vázquez, G., Pérez-Maqueo, O., Sutton, P. & Landgrave, R. (2007), 'The coasts of our world: Ecological, economic and social importance', *Ecological Economics* **63**(2–3), 254–272. doi: 10.1016/j.ecolecon.2006.10.022.

Mascagni, M. L., Siegle, E., Tessler, M. G. & Goya, S. C. (2018), 'Morphodynamics of a wave dominated embayed beach on an irregular rocky coastline', *Brazilian Journal of Oceanography* **66**(2), 172–188. doi: 10.1590/s1679-87592018005006602.

Matthews, M. W. (2011), 'A current review of empirical procedures of remote sensing in inland and near-coastal transitional waters', *International Journal of Remote Sensing* **32**(21), 6855–6899.

McCall, R., Masselink, G., Poate, T., Roelvink, J., Almeida, L., Davidson, M. & Russell, P. (2014), 'Modelling storm hydrodynamics on gravel beaches with XBeach-G', *Coastal Engineering* **91**, 231–250. doi: 10.1016/j.coastaleng.2014.06.007.

McWilliams, J. C. (2016), 'Submesoscale currents in the ocean', *Proceedings of the Royal Society of London A: Mathematical, Physical and Engineering Sciences* **472**(2189). http://rspa.royalsocietypublishing.org/content/472/2189/20160117.

Mendez, F. J. & Losada, I. J. (2004), 'An empirical model to estimate the propagation of random breaking and nonbreaking waves over vegetation fields', *Coastal Engineering* **51**(2), 103–118. doi: 10.1016/j.coastaleng.2003.11.003.

Mengual, B., Hir, P. L., Cayocca, F. & Garlan, T. (2017), 'Modelling fine sediment dynamics: Towards a common erosion law for fine sand, mud and mixtures', *Water* **9**(8). www.mdpi.com/2073-4441/9/8/564.

Menter, F. R. (1994), 'Two-equation eddy-viscosity turbulence models for engineering applications', *AIAA Journal* **32**(8), 1598–1605. doi: 10.2514/3.12149.

Middelmann-Fernandes, M. & Nielsen, O. (2009), Investing in the development of an open source two-dimensional flood modelling capability, Technical Report 2009/36, Geoscience Australia, Australia.

Millar, D., Smith, H. & Reeve, D. (2007), 'Modelling analysis of the sensitivity of shoreline change to a wave farm', *Ocean Engineering* **34**(5–6), 884–901. doi: 10.1016/j.oceaneng.2005.12.014.

Millennium Ecosystem Assessment, ed. (2005), *Ecosystems and Human Well-Being: Wetlands and Water Synthesis*, Island Press, Washington, DC.

Mofjeld, H. O. (1988), 'Depth dependence of bottom stress and quadratic drag coefficient for barotropic pressure-driven currents', *Journal of Physical Oceanography* **18**(11), 1658–1669. doi: 10.1175/1520-0485(1988)018<1658:ddobsa>2.0.co;2.

Moftakhari, H., Jay, D., Talke, S. & Schoellhamer, D. (2015), 'Estimation of historic flows and sediment loads to San Francisco Bay, 1849–2011', *Journal of Hydrology* **529**, 1247–1261. doi: 10.1016/j.jhydrol.2015.08.043.

Mohammed, Z. (2017), 'Morphodynamic modelling of sediment control groynes in a meandering river entering a reservoir'. www.ruor.uottawa. ca/handle/10393/35872.

Molemaker, M. J., McWilliams, J. C. & Dewar, W. K. (2015), 'Submesoscale instability and generation of mesoscale anticyclones near a separation of the California undercurrent', *Journal of Physical Oceanography* **45**(3), 613–629.

Möller, I. (2006), 'Quantifying saltmarsh vegetation and its effect on wave height dissipation: Results from a UK east coast saltmarsh', *Estuarine, Coastal and Shelf Science* **69**(3–4), 337–351. doi: 10.1016/j.ecss.2006.05.003.

Möller, I., Kudella, M., Rupprecht, F., Spencer, T., Paul, M., van Wesenbeeck, B. K., Wolters, G., Jensen, K., Bouma, T. J., Miranda-Lange, M. & Schimmels, S. (2014), 'Wave attenuation over coastal salt marshes under storm surge conditions', *Nature Geoscience* **7**(10), 727–731. doi: 10.1038/ngeo2251.

Monger, B. & Pershing, A. (2005), 'EAS 494: Physical Oceanography Lecture Notes', www.geo.cornell.edu/ocean/p_ocean/.

Moore, D. (1970), 'The mass transport velocity induced by free oscillations at a single frequency', *Geophysical Fluid Dynamics* **1**(1–2), 237–247.

Morang, A. & Birkemeier, W. A. (2005), *Encyclopedia of Coastal Science*, Springer, Netherlands, Dordrecht, Chapter: Depth of Closure on Sandy Coasts, pp. 374–376.

Morris, J. T., Sundareshwar, P. V., Nietch, C. T., Kjerfve, B. & Cahoon, D. R. (2002), 'Responses of Coastal Wetlands to rising sea level', *Ecology* **83**(10), 2869–2877. doi: 10.1890/0012-9658(2002)083[2869:rocwtr]2.0.co;2.

Morris, R. K. (2012), 'Managed realignment: A sediment management perspective', *Ocean and Coastal Management* **65**(Supplement C), 59–66. www.sciencedirect.com/science/article/pii/S009645691200097X.

Mudd, S. M., Howell, S. M. & Morris, J. T. (2009), 'Impact of dynamic feedbacks between sedimentation, sea-level rise, and biomass production on near-surface marsh stratigraphy and carbon accumulation', *Estuarine, Coastal and Shelf Science* **82**, 377–389.

Mulder, J. P., Hommes, S. & Horstman, E. M. (2011), 'Implementation of coastal erosion management in the Netherlands', *Ocean and Coastal Management* **54**(12), 888–897. doi: 10.1016/j.ocecoaman.2011.06.009.

NCCARF (2014), NCCARF 2008–2013: The first five years, Technical Report 122/13, National Climate Change Adaptation Research Facility, Gold Coast. www.nccarf.edu.au/sites/default/files/research_content_downloads/ NCC030-report%20FA.pdf.

Nelson, J. M., Shreve, R. L., McLean, S. R. & Drake, T. G. (1995), 'Role of near-bed turbulence structure in bed load transport and bed form mechanics', *Water Resources Research* **31**(8), 2071–2086. doi: 10.1029/95wr00976.

Nepf, H. M. (1999), 'Drag, turbulence, and diffusion in flow through emergent vegetation', *Water Resources Research* **35**(2), 479–489. doi: 10.1029/1998wr900069.

New York City Government (2019), 'Coastal storms'. www1.nyc.gov/ assets/em/downloads/pdf/hazard_mitigation/nycs_risk_landscape_chapter_ 4.1_coastalstorms.pdf. [last accessed: 08-03-2019].

Nicholls, R.J., Birkemeier, W.A. & Hallermeier, R.J. (1996), 'Application of the depth of closure concept', *Proceedings of the 25th International Conf. Coastal Engineering. ASCE, New York*, pp. 3874–3887.

Nicholls, R., Cooper, N. & Townend, I. (2007), Chapter: The management of coastal flooding and erosion. In: C. R. Thorne et al. (eds) *Future Flood and Coastal Erosion Risks*, Thomas Telford, London, pp. 392–413.

Nicholls, R., Larson, M., Capobianco, M. & Birkemeier, W. (2001), 'Depth of closure: Improving understanding and prediction', *Coastal Engineering Proceedings* **1**(26). https://icce-ojs-tamu.tdl.org/icce/index.php/icce/article/ view/5810.

Nichols, M. M. & Biggs, R. B. (1985), Estuaries. In: R. A. Davis (ed.) *Coastal Sedimentary Environments*, Springer-Verlag, Switzerland, pp. 77–186.

Niedoroda, A. W., Reed, C. W., Swift, D. J., Arato, H. & Hoyanagi, K. (1995), 'Modeling shore-normal large-scale coastal evolution', *Marine Geology* **126**(1), 181 – 199. Large-Scale Coastal Behavior. www.sciencedirect.com/science/article/pii/0025322795989617.

Nielsen, P. (2006), 'Sheet flow sediment transport under waves with acceleration skewness and boundary layer streaming', *Coastal Engineering* **53**(9), 749–758. www.sciencedirect.com/science/article/pii/S037838390 6000433.

Nordenson, G., Seavitt, C. & Yarinsky, A. (2010), *On the Water: Palisade Bay*, 1st edn, The Museum of Modern Art, New York.

Ochi, M. K. (1998), *Ocean Waves*, Cambridge University Press, Cambridge. doi: 10.1017/cbo9780511529559.

Ohlendorf, S., Müller, A., Heege, T., Cerdeira-Estrada, S. & Kobryn, H. T. (2011), Bathymetry mapping and sea floor classification using multispectral satellite data and standardized physics based data processing. In: C.R. Bostater (ed.) *'Remote Sensing of the Ocean, Sea Ice, Coastal Waters, and Large Water Regions 2011'*, Vol. 8175, 817503, SPIE, Bellingham, WA, pp. 1–9.

Olmanson, L. G., Bauer, M. E. & Brezonik, P. L. (2008), 'A 20-year Landsat water clarity census of minnesota's 10,000 lakes', *Remote Sensing of Environment* **3**, 4086–4097.

Onderka, M. & Pekarova, P. (2008), 'Retrieval of suspended particulate matter concentrations in the Danube River from Landsat ETM data', *Science of the Total Environment* **397**, 238–243.

Ortega-Rubio, A., ed. (2018), *Mexican Natural Resources Management and Biodiversity Conservation*, Springer International Publishing. doi: 10.1007/978-3-319-90584-6.

Oyama, Y., Matsushita, B., Fulkushima, K., Matsushige, T. & Imai, A. (2009), 'Application of spectral decomposition algorithm for mapping water quality in a turbid lake (Lake Kasumigaura, Japan) from Landsat TM data', *Journal of Photogrammetric Engineering and Remote Sensing* **64**, 73–85.

Packwood, A. (1983), 'The influence of beach porosity on wave uprush and backwash', *Coastal Engineering* **7**(1), 29–40. doi: 10.1016/0378-3839(83)90025-x.

Paola, C. & Voller, V. R. (2005), 'A generalized Exner equation for sediment mass balance', *Journal of Geophysical Research: Earth Surface* **110**(F4). doi: 10.1029/2004jf000274.

Parker, G. (2008), Transport of gravel and sediment mixtures. In: *'Sedimentation Engineering'*, American Society of Civil Engineers, pp. 165–251. doi: 10.1061/9780784408148.ch03.

Pauline, W., M., M. K., Martin, J., Thierry, S., Shin, T., Erik, A. J., Marzia, R., Dale, C., Vicki, F. & Rochelle, W. (2015), 'A new digital bathymetric model of the world's oceans', *Earth and Space Science* **2**(8), 331–345. https://agupubs.onlinelibrary.wiley.com/doi/abs/10.1002/2015EA000107.

Pawlowicz, R., Beardsley, B. & Lentz, S. (2002), 'Classical tidal harmonic analysis including error estimates in MATLAB using TTIDE', *Computers and Geosciences* **28**, 929–937.

Pedrozo-Acuña, A., Torres-Freyermuth, A., Zou, Q., Hsu, T.-J. & Reeve, D. E. (2010), 'Diagnostic investigation of impulsive pressures induced by plunging breakers impinging on gravel beaches', *Coastal Engineering* **57**(3), 252–266. doi: 10.1016/j.coastaleng.2009.09.010.

Peregrine, D. H. (1972), Equations for water waves and the approximation behind them. In: *'Waves on Beaches and Resulting Sediment Transport'*, Elsevier, pp. 95–121. doi: 10.1016/b978-0-12-493250-0.50007-2.

Perillo, G. (1995), *Definition and Geomorphologic Classifications of Estuaries*, Elsevier, Netherlands, Amsterdam, pp. 17–47.

Perillo, G. M. E. & Piccolo, M. C. (2011), *Global Variability in Estuaries and Coastal Settings*, Vol. 1, Academic Press, Waltham, pp. 7–36.

Persson, P. & Strang, G. (2004), 'A simple mesh generator in MATLAB', *SIAM Review* **46**(2), 329–345.

Phillips, N. A. (1957), 'A coordinate system having some special advantages for numerical forecasting', *Journal of Meteorology* **14**(2), 184–185. doi: 10.1175/1520-0469(1957)014<0184:acshss>2.0.co;2.

Phillips, O. M. (1977), *The Dynamics of the Upper Ocean*, Cambridge University Press, London.

Popinet, S. (2003), 'Gerris: A tree-based adaptive solver for the incompressible euler equations in complex geometries', *Journal of Computational Physics* **190**(2), 572–600. doi: 10.1016/s0021-9991(03)00298-5.

Popinet, S. & Rickard, G. (2007), 'A tree-based solver for adaptive ocean modelling', *Ocean Modelling* **16**(3), 224–249. www.sciencedirect.com/science/article/pii/S1463500306000989.

Pourzangbar, A., Saber, A., Yeganeh-Bakhtiary, A. & Ahari, L. R. (2017), 'Predicting scour depth at seawalls using GP and ANNs', *Journal of Hydroinformatics* **19**(3), 349–363. doi: 10.2166/hydro.2017.125.

Prandle, D. & Wolf, J. (1978), Surge-tide interaction in the southern north sea. In: J. C. Nihoul, (ed.) *'Hydrodynamics of Estuaries and Fjords'*, Vol. 23, *Elsevier Oceanography Series*, Elsevier, pp. 161–185. www.sciencedirect.com/science/article/pii/S0422989408712777.

Pritchard, D. (1952), Estuarine hydrography. In: *'Advances in Geophysics Volume 1'*, Elsevier, pp. 243–280. doi: 10.1016/s0065-2687(08)60208-3.

Pugh, D. T. (1996), *Tides, Surges and Mean Sea-Level*, John Wiley and Sons Ltd, Hoboken, NJ.

Quataert, E., Storlazzi, C., van Rooijen, A., Cheriton, O. & Van Dongeren, A. (2015), 'The influence of coral reefs and climate change on wave-driven flooding of tropical coastlines', *Geophysical Research Letters* **42**(15), 6407–6415. doi: 10.1002/2015gl064861.

Rafiee, A., Pistani, F. & Thiagarajan, K. (2010), 'Study of liquid sloshing: Numerical and experimental approach', *Computational Mechanics* **47**(1), 65–75. doi: 10.1007/s00466-010-0529-6.

Raupach, M. R., Antonia, R. A. & Rajagopalan, S. S. (1991), 'Rough-wall turbulent boundary layers', *ASME Applied Mechanics Reviews* **44**(1), 1–25.

Razak, M. S. A. & Nor, N. A. Z. M. (2018), 'XBeach process-based modelling of coastal morphological features near breakwater', *MATEC Web of Conferences* **203**, 01007. doi: 10.1051/matecconf/201820301007.

Reed, D. J. (1995), 'The response of coastal marshes to sea-level rise: Survival or submergence?' *Earth Surface Processes and Landforms* **20**(1), 39–48. doi: 10.1002/esp.3290200105.

Reeve, D., Chadwick, A. & Fleming, C. (2018), *Coastal Engineering: Theory, Processes, and Design Practice*, 3rd edn, CRC Press, Boca Raton, FL.

Reeve, D., Chen, Y., Pan, S., Magar, V., Simmonds, D. & Zacharioudaki, A. (2011), 'An investigation of the impacts of climate change on wave energy generation: The wave hub, cornwall, UK', *Renewable Energy* **36**(9), 2404–2413. doi: 10.1016/j.renene.2011.02.020.

Reeve, D. E., Horrillo-Caraballo, J. M. & Magar, V. (2008), 'Statistical analysis and forecasts of long-term sandbank evolution at Great Yarmouth, UK', *Estuarine, Coastal and Shelf Science* **79**(3), 387–399. www.sciencedirect.com/science/article/pii/S0272771408001807.

Reeve, D. E., Karunarathna, H., Pan, S., Horrillo-Caraballo, J. M., Różyński, G. & Ranasinghe, R. (2016), 'Data-driven and hybrid coastal morphological prediction methods for mesoscale forecasting', *Geomorphology* **256**, 49–67. doi: 10.1016/j.geomorph.2015.10.016.

Reisinger, A., Kitching, R., Chiew, F., Hughes, L., Newton, P., Schuster, S., Tait, A. & Whetton, P. (2014), Chapter 25, Contribution of Working Group II to the Fifth Assessment Report of the Intergovernmental Panel on Climate Change. In: V. R. Barros et al. (eds.) *Climate Change 2014: Impacts, Adaptation, and Vulnerability. Part B: Regional Aspects*, Cambridge University Press, Cambridge, UK and New York, p. 688.

Reisinger, A., Lawrence, J., Hart, G. & Chapman, R. (2015), Chapter 13, *From Coping to Resilience: The Role of Managed Retreat in Highly Developed Coastal Regions of New Zealand*, CRC Press, Boca Raton, FL.

Reniers, A. J. H. M. (2004), 'Morphodynamic modeling of an embayed beach under wave group forcing', *Journal of Geophysical Research* **109**(C1). doi: 10.1029/2002jc001586.

Reynaud, J.-Y. & Dalrymple, R. W. (2011), Shallow-marine tidal deposits. In: *'Principles of Tidal Sedimentology'*, Springer, Netherlands, pp. 335–369. doi: 10.1007/9789400701236_13.

Rinaldo, A., Fagherazzi, S., Lanzoni, S., Marani, M. & Dietrich, W. E. (1999), 'Tidal networks 3. landscape-forming discharges and studies in empirical geomorphic relationships', *Water Resources Research* **35**(12), 3919–3929.

Ritchie, J., Zimba, P. & Everitt, J. (2003), 'Remote sensing techniques to assess water quality', *Photogrammetric Engineering and Remote Sensing* **69**, 695–704.

Rivier, A., Bennis, A.-C., Pinon, G., Magar, V. & Gross, M. (2016), 'Parameterization of wind turbine impacts on hydrodynamics and sediment transport', *Ocean Dynamics* **66**(10), 1285–1299. doi: 10.1007/s10236-016-0983-6.

Roelvink, D. & Reniers, A. (2011), Modeling approaches. In: *'Advances in Coastal and Ocean Engineering'*, World Scientific, pp. 145–167.

Roelvink, D. & Reniers, A. (2012), Chapter 6, *Modeling approaches*. In: P. L. F Liu (ed.) *Advances in Coastal and Ocean Engineering*, World Scientific, Singapore, Vol. 12 , pp. 145–167.

Roelvink, D., Reniers, A., Van Dongeren, A., van Thiel de Vries, J., McCall, R. & Lescinski, J. (2009), 'Modelling storm impacts on beaches, dunes and barrier islands', *Coastal Engineering* **56**(11–12), 1133–1152. doi: 10.1016/j.coastaleng.2009.08.006.

Roelvink, D., Van Dongeren, A., McCall, R., Hoonhout, B., van Rooijen, A., van Geer, P., de Vet, L., Nederhoff, K. & Quataert, E. (2015), XBeach Technical Reference: Kingsday Release, Technical Report, UNESCO-IHE Institute of Water Education and Delft University of Technology, Deltares.

Roelvink, J. A. & Stive, M. J. F. (1989), 'Bar-generating cross-shore flow mechanisms on a beach', *Journal of Geophysical Research* **94**(C4), 4785. doi: 10.1029/jc094ic04p04785.

Roelvink, J. A. & Walstra, D. J. R. (2004), Keeping it simple by using complex models, *Advances in Hydro-science and Engineering* **6**, 1–11, University of Mississippi, Oxford, MS.

Ross, D. A. (1995), *Introduction to Oceanography*, Harper Collins, New York.

Ross, L., Huguenard, K. & Sottolichio, A. (2019), 'Intratidal and fortnightly variability of vertical mixing in a macrotidal estuary: The Gironde', *Journal of Geophysical Research: Oceans* **124**(4), 2641–2659. doi: 10.1029/2018jc014456.

Różyński, G. (2005), 'Long-term shoreline response of a nontidal, barred coast', *Coastal Engineering* **52**(1), 79–91. doi: 10.1016/j.coastaleng.2004.09.007.

Różyński, G., Larson, M. & Pruszak, Z. (2001), 'Forced and self-organized shoreline response for a beach in the southern Baltic Sea determined through singular spectrum analysis', *Coastal Engineering* **43**(1), 41–58. doi: 10.1016/s0378-3839(01)00005-9.

Ruggiero, P., Kaminsky, G. M., Gelfenbaum, G. & Voigt, B. (2005), 'Seasonal to interannual morphodynamics along a high-energy dissipative littoral cell', *Journal of Coastal Research* **213**, 553–578. doi: 10.2112/03-0029.1.

Salim, S., Pattiaratchi, C., Tinoco, R., Coco, G., Hetzel, Y., Wijeratne, S. & Jayaratne, R. (2017), 'The influence of turbulent bursting on sediment resuspension under unidirectional currents', *Earth Surface Dynamics* **5**(3), 399–415. doi: 10.5194/esurf-5-399-2017.

Salisbury, M. B. & Hagen, S. C. (2007), 'The effect of tidal inlets on open coast storm surge hydrographs', *Coastal Engineering* **54**(5), 377–391. doi: 10.1016/j.coastaleng.2006.10.002.

Sanay, R., Voulgaris, G. & Warner, J. C. (2007), 'Tidal asymmetry and residual circulation over linear sandbanks and their implication on sediment transport: A process-oriented numerical study', *Journal of Geophysical Research: Oceans* **112**(C12). https://agupubs.onlinelibrary.wiley.com/doi/abs/10.1029/2007JC004101.

Sand Motor (2019). www.dezandmotor.nl/en/the-sand-motor/introduction/.

Schoellhamer, D. (2011), 'Sudden clearing of estuarine waters upon crossing the threshold from transport to supply regulation of sediment transport as an erodible sediment pool is depleted: San Francisco Bay, 1999', *Estuaries Coasts* **34**(5), 885–899.

Schoellhamer, D. H., Ganju, N. K. & Shellenbarger, G. G. (n.d.), *Sediment Transport in San Pablo Bay*, U.S. Army Corps of Engineers, San Francisco, CA.

Seminara, G. & Blondeaux, P., eds (2001), *River, Coastal and Estuarine Morphodynamics*, Springer-Verlag, Berlin, Heidelberg.

Shields, A. F. (1936), (Application of Similarity Principles and Turbulence Research to Bed-Load Movement), PhD thesis, Prussian Research Institute for Hydraulic Engineering and Shipbuilding, Berlin: Germany. Translated from: 'Anwendung der Aehnlichkeitsmechanilr und der Turbulenzforschung ad die Geschiebe - bewegung', Mitteilungen der Preussischen Versuchsanstalt fur Nassexbau nd Schiffbau, Berlin, 1936, by W. P. Ott and J. C. van Uchelen.

Shih, T. H., Zhu, J. & Lumley, J. L. (1996), 'Calculation of wall-bounded complex flows and free shear flows', *International Journal for*

Numerical Methods in Fluids **23**(11), 1133–1144. doi: 10.1002/(sici)1097-0363(19961215)23:11<1133::aid-fld456>3.0.co;2-a.

Sleath, J. (1999), 'Conditions for plug formation in oscillatory flow', *Continental Shelf Research* **19**(13), 1643–1664. www.sciencedirect.com/science/article/pii/S027843439800096X.

Sleath, J. F. A. (1982), 'The suspension of sand by waves', *Journal of Hydraulic Research* **20**(5), 439–452.

Smith, R. (1976), 'Longitudinal dispersion of a buoyant contaminant in a shallow channel', *Journal of Fluid Mechanics* **78**(04), 677. doi: 10.1017/s0022112076002681.

Smith, R. (1977), 'Long-term dispersion of contaminants in small estuaries', *Journal of Fluid Mechanics* **82**(1), 129–146. doi: 10.1017/s002211 2077000561.

Smith, S. D. (1988), 'Coefficients for sea surface wind stress, heat flux, and wind profiles as a function of wind speed and temperature', *Journal of Geophysical Research: Oceans* **93**(C12), 15467–15472.

Smith, S. D. & Banke, E. G. (1975), 'Variation of the sea surface drag coefficient with wind speed', *Quarterly Journal of the Royal Meteorological Society* **101**(429), 665–673. doi: 10.1002/qj.49710142920.

Solecki, W., Rosenzweig, C., Gornitz, V., Horton, R., Major, D. C., Patrick, L. & Zimmerman, R. (2015), Chapter 5: Climate change and infrastructure adaptation in coastal New York City. In: B. Glavovic, R. Kay, M. Kelly, & A. Travers (eds.) *Climate Change and the Coast. Building Resilient Communities*, Taylor & Francis Group, CRC Press, London, pp. 125–146.

Soloviev, A. & Lukas, R. (1996), 'Observation of spatial variability of diurnal thermocline and rain-formed halocline in the western Pacific warm pool', *Journal of Physical Oceanography* **26**(11), 2529–2538. doi: 10.1175/1520-0485(1996)026<2529:oosvod>2.0.co;2.

Soulsby, R. (1997), *Dynamics of Marine Sands.*, Thomas Telford Publications, London, pp. 1–173.

Soulsby, R. L. & Damgaard, J. S. (2005), 'Bedload sediment transport in coastal waters', *Coastal Engineering* **52**(8), 673–689. doi: 10.1016/j.coastaleng.2005.04.003.

South Ponte Vedra-Vilano Beach Preservation Association, I. S. (2015), 'Protecting the beaches that we love'. www.spv-vilano.com.

Southard, J. (2006), Lecture 9: Threshold of movement. In: *'Introduction to Fluid Motions, Sediment Transport, and Current-Generated Sedimentary Structures. Fall 2006 — MIT Course No. 12.090'*, MIT Open-CourseWare. License: Creative Commons BY-NC-SA., Cambridge MA. https://ocw.mit.edu.

Southgate, H. N., Wijnberg, K. M., Larson, M., Capobianco, M. & Jansen, H. (2003), 'Analysis of field data of coastal morphological evolution over yearly and decadal timescales. Part 2: Non-linear techniques', *Journal of Coastal Research* **19**(4), 776–789. www.jstor.org/stable/4299220.

Srinivasan, K., McWilliams, J. C. & Molemaker, J. (2017), Submesoscale topographic wakes, *In '21st Conference on Atmospheric and Oceanic Fluid Dynamics and the 19th Conference on Middle Atmosphere Book of Abstracts'*, American Meteorological Society, Boston, MA. https://ams.confex.com/ams/21Fluid19Middle/webprogram/Paper 319654.html.

Stive, M. J. F. (2003), Advances in morphodynamics of coasts and lagoons. *In 'Proceedings of the International Conference on Estuaries and Coasts 2003'*, Hangzhou, China.

Stokes, G. G. (1880), 'On the theory of oscillatory waves', *Mathematical and Physical Papers* **1**, 197–229. https://archive.org/details/ mathphyspapers01stokrich/page/n197.

Strahler, A. N. (1952), 'Hypsometric (area-altitude) analysis of erosional topography, *Geological Society of America Bulletin* **63**(11), 1117. doi: 10.1130/0016-7606(1952)63[1117:haaoet]2.0.co;2.

Sumer, B., Kozakiewicz, A., Fredsoe, J. & Deigaard, R. (1996), 'Velocity and concentration profiles in sheet-flow layer of movable bed', *Journal of Hydraulic Engineering - ASCE* **122**(10), 549–558.

Sutherland, J., Peet, A. & Soulsby, R. (2004), 'Evaluating the performance of morphological models', *Coastal Engineering* **51**(8–9), 917–939. doi: 10.1016/j.coastaleng.2004.07.015.

Svendsen, I. (1984), 'Mass flux and undertow in a surf zone', *Coastal Engineering* **8**(4), 347–365. doi: 10.1016/0378-3839(84)90030-9.

Syvitski, J. M. P., Harvey, N., Wolanski, E., Burnett, W. C., Perillo, G. M. E., Gornitz, V., Bokuniewicz, H., Huettel, M., Moore, W. S., Saito, Y., Taniguchi, M., Hesp, P., Yim, W. W.-S., Salisbury, J., Campbell, J., Snoussi, M., Haida, S., Arthurton, R. & Gao, S. (2005), *Dynamics of the Coastal Zone*, Springer, Germany, Berlin, pp. 39–94.

Talke, S. A., de Swart, H. & de Jonge, V. N. (2009), 'An idealized model and systematic process study of oxygen depletion in highly turbid estuaries', *Estuaries and Coasts* **32**(4), 602–620. doi: 10.1007/s12237-009-9171-y.

Temmerman, S., Govers, G., Meire, P. & Wartel, S. (2003), 'Modelling long-term tidal marsh growth under changing tidal conditions and suspended sediment concentrations, Scheldt estuary, Belgium', *Marine Geology* **193**(1–2), 151–169. doi: 10.1016/s0025-3227(02)00642-4.

Tennekes, H. & Lumley, J. L. (1972), *A First Course in Turbulence*, MIT Press, Cambridge, MA. ISBN: 9780262536301.

Terrile, E., Reniers, A. J., Stive, M. J., Tromp, M. & Verhagen, H. J. (2006), 'Incipient motion of coarse particles under regular shoaling waves', *Coastal Engineering* **53**(1), 81–92. www.sciencedirect.com/science/article/pii/S0378383905001249.

Thomas, L. N., Tandon, A. & Mahadevan, A. (2013), *Submesoscale Processes and Dynamics*, American Geophysical Union, Washington, DC, pp. 17–38.

Thurman, H. V. (1994), *Introductory Oceanography*, 7th edn, Macmillan, New York.

Tolman, H. L. (1991), 'A third-generation model for wind waves on slowly varying, unsteady, and inhomogeneous depths and currents', *Journal of Physical Oceanography* **21**(6), 782–797.

Tonnon, P. K., Huisman, B. J. A., Stam, G. N. & Van Rijn, L. C. (2018), 'Numerical modelling of erosion rates, life span and maintenance volumes of mega nourishments', *Coastal Engineering* **131**, 51–69.

Toro, E. F. (1989), 'A weighted average flux method for hyperbolic conservation laws', *Proceedings of the Royal Society A: Mathematical, Physical and Engineering Sciences* **423**(1865), 401–418. doi: 10.1098/rspa.1989.0062.

Toro, E. F. (1992), 'Riemann problems and the WAF method for solving the two-dimensional shallow water equations', *Philosophical Transactions of the Royal Society of London. Series A: Physical and Engineering Sciences* **338**(1649), 43–68. doi: 10.1098/rsta.1992.0002.

Uncles, R. (1981), 'A note on tidal asymmetry in the Severn Estuary', *Estuarine, Coastal and Shelf Science* **13**(4), 419–432. www.sciencedirect.com/science/article/pii/S0302352481800382.

Uncles, R. J. & Mitchell, S. B. (2017), Estuarine and coastal hydrography and sediment transport. In: R. J. Uncles & S. B. Mitchell, (eds). '*Estuarine and Coastal Hydrography and Sediment Transport*', Cambridge University Press, pp. 1–34. doi: 10.1017/9781139644426.002.

Uncles, R. & Jordan, M. (1980), 'A one-dimensional representation of residual currents in the Severn Estuary and associated observations', *Estuarine and Coastal Marine Science* **10**(1), 39–60. www.sciencedirect.com/science/article/pii/S030235248080048X.

US EPA (n.d.). www.epa.gov/sfbay-delta/about-watershed. [last accessed: 09-03-2019].

Valle-Levinson, A. (2010), *Contemporary Issues in Estuarine Physics*, Cambridge University Press, Cambridge.

Van der A, D., O'Donoghue, T. & Ribberink, J. S. (2010), 'Measurements of sheet flow transport in acceleration-skewed oscillatory flow and comparison with practical formulations', *Coastal Engineering* **57**(3), 331–342.

Van der Wegen, M. & Jaffe, B. (2013), 'Towards a probabilistic assessment of process-based, morphodynamic models', *Coastal Engineering* **75**, 52–63. doi: 10.1016/j.coastaleng.2013.01.009.

Van der Wegen, M. & Jaffe, B. E. (2014), 'Processes governing decadal scale depositional narrowing of the major tidal channel in San Pablo Bay, California, USA', *Journal of Geophysical Research: Earth Surface* **119**(5), 1136–1154. doi: 10.1002/2013jf002824.

Van der Wegen, M. & Roelvink, J. A. (2008), 'Long-term morphodynamic evolution of a tidal embayment using a two-dimensional, process-based model', *Journal of Geophysical Research* **113**(C3). doi: 10.1029/2006jc003983.

Van der Wegen, M., Jaffe, B. E. & Roelvink, J. A. (2011), 'Process-based, morphodynamic hindcast of decadal deposition patterns in San Pablo Bay, California, 1856–1887', *Journal of Geophysical Research: Earth Surface* **116**(F2). doi: 10.1029/2009jf001614.

Van Drie, R., Milevski, P. & Simon, M. (2011), Validation of a 2-D hydraulic model -ANUGA, to undertake hydrologic analysis. *'In Conference of 34th Biennial Congress of the International Association of Hydraulic Engineering and Research'.* Available from: www.researchgate.net/ publication/309682409_Validation_of_a_2-D_Hydraulic_Model_-ANUGA_to_ undertake_Hydrologic_Analysis. [accessed: 19-02-2019].

Van der Wegen, M., Dastgheib, A., Jaffe, B. E. & Roelvink, D. (2010), 'Bed composition generation for morphodynamic modeling: Case study of San Pablo Bay in California, USA', *Ocean Dynamics* **61**(2–3), 173–186. doi: 10.1007/s10236-010-0314-2.

Van der Wegen, M., Jaffe, B., Foxgrover, A. & Roelvink, D. (2016), 'Mudat morphodynamics and the impact of sea level rise in South San Francisco Bay', *Estuaries and Coasts* **40**(1), 37–49. doi: 10.1007/s12237-016-0129-6.

Van der Werf, J., Magar, V., Malarkey, J., Guizien, K. & O'Donoghue, T. (2008), '2DV modelling of sediment transport processes over fullscale ripples in regular asymmetric oscillatory flow', *Continental Shelf Research* **28**(8), 1040–1056. www.sciencedirect.com/science/article/pii/ S0278434308000459.

Van der Westhuysen, A. J., Van Dongeren, A. R., Groeneweg, J., Van Vledder, G. P., Peters, H., Gautier, C. & Van Nieuwkoop, J. C. C. (2012), 'Improvements in spectral wave modeling in tidal inlet seas', *Journal of Geophysical Research: Oceans* **117**(C11). doi: 10.1029/2011jc007837.

Van Dongeren, A., Lowe, R., Pomeroy, A., Minh Trang, D., Roelvink, D., Symonds, G. & Ranasinghe, R. (2013), 'Numerical modeling of low-frequency wave dynamics over a fringing coral reef', *Coastal Engineering* **73**, 178–190. doi: 10.1016/j.coastaleng.2012.11.004.

Van Dongeren, A., Plant, N., Cohen, A., Roelvink, D., Haller, M. C. & Catalán, P. (2008), 'Beach wizard: Nearshore bathymetry estimation through assimilation of model computations and remote observations', *Coastal Engineering* **55**(12), 1016–1027. doi: 10.1016/j.coastaleng.2008.04.011.

Van Duin, M., Wiersma, N., Walstra, D., Van Rijn, L. & Stive, M. (2004), 'Nourishing the shoreface: observations and hindcasting of the Egmond case, The Netherlands', *Coastal Engineering* **51**(8–9), 813–837. doi: 10.1016/j.coastaleng.2004.07.011.

Van Hooft, J. A., Popinet, S. & Van de Wiel, B. (2018), 'Adaptive Cartesian meshes for atmospheric single-column models: a study using Basilisk 18-02-16', *Geoscience Model Development* **11**, 4727—4738, doi:10.5194/gmd-11-4727-2018.

Van Koningsveld, M. & Mulder, J. P. M. (2004), 'Sustainable coastal policy developments in the Netherlands. A systematic approach revealed', *Journal of Coastal Research* **20**(2), 375–385.

Van Koningsveld, M., Davidson, M. A. & Huntley, D. A. (2005), 'Matching science with coastal management needs: the search for appropriate coastal state indicators', *Journal of Coastal Research* **21**(3), 399–411.

Van Maren, D. S. (2009), 'Background theory cohesive-fine sediment-mud'. http://oss.deltares.nl/web/delft3d/community-wiki. [last accessed: 25-03-2018].

Van Maren, D. S. & Winterwerp, J. (2013), 'The role of flow asymmetry and mud properties on tidal flat sedimentation', *Continental Shelf Research* **60**, S71 – S84. Hydrodynamics and sedimentation on mesotidal sand- and mudflats. www.sciencedirect.com/science/article/pii/S0278434312001951.

Van Rijn, L. C. (1993), *Principles of Sediment Transport in Rivers Estuaries and Coastal Seas*, Aqua Publications, Blokzijl, Netherlands. pp. 1.1–2.1.

Van Rijn, L. C. (2007), 'Unified view of sediment transport by currents and waves. I: Initiation of motion, bed roughness, and bed-load transport',

Journal of Hydraulic Engineering **133**(6), 649–667. doi: 10.1061/(asce)0733-9429(2007)133:6(649).

Van Rijn, L. C., Walstra, D. J. R. & van Ormondt, M. (2004), Description of transpor 2004 (tr2004) and implementation in DELFT3D-online, Technical Report Z3748, Delft Hydraulics, Delft, The Netherlands.

Van Rijn, L., Soulsby, R., Hoekstra, P. & Davies, A. G., eds (2005), *SAND-PIT: Sand Transport and Morphology of Offshore Sand Mining Pits. Process knowledge and guidelines for coastal management. End document May 2005, EC Framework V Project No. EVK3-2001-00056*, Aqua Publications, The Netherlands.

Vautard, R. & Ghil, M. (1989), 'Singular spectrum analysis in nonlinear dynamics, with applications to paleoclimatic time series', *Physica D: Nonlinear Phenomena* **35**(3), 395–424. doi: 10.1016/0167-2789(89)90077-8.

Veeramony, J. & Svendsen, I. (2000), 'The flow in surf-zone waves', *Coastal Engineering* **39**(2–4), 93–122. doi: 10.1016/s0378-3839(99)00058-7.

Vittori, G. & Verzico, R. (1998), 'Direct simulation of transition in an oscillatory boundary layer', *Journal of Fluid Mechanics* **371**, 207–232. doi: 10.1017/s002211209800216x.

Vreugdenhil, C. (2006), 'Appropriate models and uncertainties', *Coastal Engineering* **53**(2–3), 303–310. doi: 10.1016/j.coastaleng.2005.10.017.

WAMDI Group. (1988), 'The WAM model: A third generation ocean wave prediction model', *Journal of Physical Oceanography* **18**(12), 1775–1810.

Warner, J. C., Sherwood, C. R., Signell, R. P., Harris, C. K. & Arango, H. G. (2008), 'Development of a three-dimensional, regional, coupled wave, current, and sediment-transport model', *Computers and Geosciences* **34**(10), 1284–1306. Predictive Modeling in Sediment Transport and Stratigraphy. www.sciencedirect.com/science/article/pii/S0098300408000563.

Waterhouse, A. F., Valle-Levinson, A. & Winant, C. D. (2011), 'Tides in a system of connected estuaries', *Journal of Physical Oceanography* **41**(5), 946–959.

Wavewatch III Development Group (2016), User manual and system documentation of Wavewatch III version 5.16, Technical Report Tech. Note 329, NOAA/NWS/NCEP/MMAB, College Park, MD.

Weather2 (2019), 'Wind, wave and weather reports, forecasts and statistics worldwide'. www.myweather2.com/City-Town/United-States-Of-America/ California/San-Pablo-Bay/climate-profile.aspx?month=12. [last accessed: 10-03-2019].

Web of Science (2019), 'Offshore wind publication topic, between 1980 and 2019'. www.webofknowledge.com.

Wei, G., Kirby, J. T., Grilli, S. T. & Subramanya, R. (1995), 'A fully nonlinear Boussinesq model for surface waves. Part 1. Highly nonlinear unsteady waves', *Journal of Fluid Mechanics* **294**, 71–92. doi: 10.1017/s0022112095002813.

Westerink, J. J. (2003), 'Development of the governing equations at a point for a turbulent time-averaged continuum'. https://coast.nd.edu/jjwteach/ www/www/344/PsNotes/topic1_4.pdf.

Whitehouse, R., Bassoullet, P., Dyer, K., Mitchener, H. & Roberts, W. (2000), 'The influence of bedforms on flow and sediment transport over intertidal mudflats', *Continental Shelf Research* **20**(10–11), 1099–1124. doi: 10.1016/s0278-4343(00)00014-5.

Wijnberg, K. & Holman, R. (2007), Video-observations of shoreward propagating accretionary waves. In: C. Dohmen - Janssen & S. Hulscher, (eds). *'River, Coastal and Estuarine Morphodynamics, RCEM 2007, 17–21 September 2007, Enschede, the Netherlands, Vol II'*, Taylor and Francis, pp. 737–743.

Wijnberg, K. M. & Terwindt, J. H. (1995), 'Extracting decadal morphological behaviour from high-resolution, long-term bathymetric surveys along the holland coast using eigenfunction analysis', *Marine Geology* **126**(1–4), 301–330. doi: 10.1016/0025-3227(95)00084-c.

Wilcox, D. C. (1993), *Turbulence Modelling for CFD*, DCW Industries, La Cañada.

Williams, J. J., Carling, P. A., Amos, C. L. & Thompson, C. (2008), 'Field investigation of ridge–runnel dynamics on an intertidal mudflat', *Estuarine, Coastal and Shelf Science* **79**(2), 213–229. doi: 10.1016/j.ecss.2008.04.001.

Willmott, C. J. (1981), 'On the validation of models', *Physical Geography* **2**(2), 184–194. doi: 10.1080/02723646.1981.10642213.

Winant, C. D., Inman, D. L. & Nordstrom, C. E. (1975), 'Description of seasonal beach changes using empirical eigenfunctions', *Journal of Geophysical Research* **80**(15), 1979–1986. doi: 10.1029/jc080i015 p01979.

Windfinder (2019), 'Coastal storms'. www.windfinder.com/windstatistics/ davis_point_san_pablo_bay. [last accessed: 10-03-2019].

Winter, C. (2006), 'Meso-scale morphodynamics of the Eider estuary: analysis and numerical modelling', *Journal of Coastal Research, SI 39 (Proceedings of the 8th International Coastal Symposium)*, Santa Catarina, Brazil, pp. 498–503.

Winter, C. (2011), Observations and Modelling of Morphodynamics in Sandy Coastal Environments, PhD thesis, University of Bremen.

Wright, L. & Short, A. (1984), 'Morphodynamic variability of surf zones and beaches: A synthesis', *Marine Geology* **56**(1–4), 93–118. doi: 10.1016/0025-3227(84)90008-2.

Xian, G. & Weng, Q., eds (2015), *Remote Sensing Applications for the Urban Environment*, Remote Sensing Applications Series, 1st edn, CRC Press, Inc., Boca Raton, FL.

Yacobi, Y. Z., Gitelson, A. & Mayo, M. (1995), 'Remote sensing of Chlorophyll in Lake Kinneret using high spectral-resolution radiometer and Landsat TM: Spectral features of reflectance and algorithm development', *Journal of Plankton Research* **17**, 2155–2217.

Zhu, S.-P. & Mitchell, L. (2009), 'Diffraction of ocean waves around a hollow cylindrical shell structure', *Wave Motion* **46**(1), 78–88. doi: 10.1016/j.wavemoti.2008.09.001.

Zijlema, M., Stelling, G. & Smit, P. (2011), 'Swash: An operational public domain code for simulating wave fields and rapidly varied flows in coastal waters', *Coastal Engineering* **58**(10), 992–1012. www.sciencedirect.com/science/article/pii/S0378383911000974.

Zoppou, C. & Roberts, S. (1999), 'Catastrophic collapse of water supply reservoirs in urban areas', *Journal of Hydraulic Engineering* **125**(7), 686–695.

Index